Industrial and Applied Mathematics

Industrial and Applied Mathematics, a Scopus-indexed book series, publishes high-quality research-level monographs, lecture notes and contributed volumes focusing on areas where mathematics is used in a fundamental way, such as industrial mathematics, bio-mathematics, financial mathematics, applied statistics, operations research and computer science.

Lyudmila Alexeyeva · Aigulim Bayegizova

Generalized Functions Method in Boundary Value Problems for Wave Equations

Springer

Lyudmila Alexeyeva
Department of Mathematical Physics and Modeling
Institute of Mathematics and Mathematical Modeling
Almaty, Kazakhstan

Aigulim Bayegizova
Department of Information Security
L. N. Gumilyov Eurasian National University
Astana, Kazakhstan

ISSN 2364-6837 ISSN 2364-6845 (electronic)
Industrial and Applied Mathematics
ISBN 978-981-95-2801-1 ISBN 978-981-95-2802-8 (eBook)
https://doi.org/10.1007/978-981-95-2802-8

Mathematics Subject Classification: 81Q05, 35A08, 35L15, 46F10, 35L10, 78A25, 35L20, 46F12

This Springer imprint is published by the registered company Springer Nature Singapore Pte Ltd.
The registered company address is: 152 Beach Road, #21-01/04 Gateway East, Singapore 189721, Singapore

Foreword

The study of wave propagation processes in bodies and environments is associated with the solution of boundary value problems for differential equations and systems that are hyperbolic or have a mixed type. Therefore, it is very important to construct effective methods for solving such equations and systems for areas with arbitrary boundary geometry and various types of boundary conditions.

A feature of hyperbolic equations is the existence of non-differentiable solutions, which is associated with shock waves. This makes it impossible to apply standard analytical methods for solving boundary value problems that use the properties of smoothness of solutions. Numerical methods based on difference approximation of equations, which also require the corresponding smoothness of solutions, often become ineffective.

One of the effective methods developed for solving boundary value problems of mathematical physics is the generalized functions method (GFM), which the authors of this monograph have been developing for many years to solve boundary value problems of mathematical physics, mechanics of deformable solids, and electrodynamics. This monograph presents in detail the GFM for solving boundary value problems for classical hyperbolic equations of mathematical physics and electrodynamics.

The monograph will be very useful for master's and doctoral students in mechanics, mathematics, and physics departments of universities, who can use this monograph to study new modern methods for constructing solutions to equations of mathematical and theoretical physics and boundary value problems. Its results and calculation formulas will be applicable in theoretical physics, radio physics, electrodynamics, geophysics, dynamics of structures, and constructions.

Prof. Nazarbay Bliev
The National Academy of Sciences
Almaty, Kazakhstan

Preface

The monograph presents the method of generalized functions and the method of boundary integral equations for solving nonstationary and stationary boundary value problems for classical hyperbolic equations of mathematical physics and electrodynamics: the wave equation, the Klein-Gordon equation, the Schrödinger equation, the Maxwell system of equations, etc. in spaces of dimension 1, 2, 3. The apparatus of the theory of generalized functions is absolutely necessary for solving hyperbolic equations and systems described by pseudo-differential operators. It allows one to describe and study processes accompanied by shock waves, which is often simply impossible within the framework of the classical theory of differential equations. Generalized solutions of the considered boundary value problems, their regular integral representations and resolving singular boundary integral equations have been constructed, which belong to a new class of boundary integral equations, which can become the subject of a separate study in the field of functional analysis and function theory.

The monograph can be recommended to master's students and research scholars of mechanics and mathematics to study the methods of constructing solutions of equations of mathematical physics and boundary value problems. The authors hope that this monograph will help the reader in reading the works presented in the bibliography related to solving more complex boundary value problems for hyperbolic systems of equations describing the wave dynamics of elastic and electromagnetic media, and most importantly in their practical work related to the study of dynamic wave processes in solid bodies and continuous media.

The monograph was written with the financial support of the Science Committee of the Republic of Kazakhstan under the Republican Program BR20281002 "Fundamental Research in Mathematics and Mathematical Modeling" and Grant AP19674789 (2023–2025).

Almaty, Kazakhstan — Lyudmila Alexeyeva
Astana, Kazakhstan — Aigulim Bayegizova

Introduction

The solution of many problems of acoustics, hydromechanics, elasticity theory, field theory, and other branches of physics and mechanics is associated with the solution of boundary value problems for wave equations describing the processes of wave propagation in homogeneous isotropic media that are hyperbolic. Therefore, it is very important to construct effective methods for solving hyperbolic equations and systems for regions with arbitrary boundary geometry and a diverse type of boundary conditions.

The theory of hyperbolic equations has an extensive bibliography. A great contribution to its development was made by many prominent scientists: J. L. d'Alembert, G. F. B. Riemann, G. R. Kirchhoff, O. L. Cauchy, J. G. Darboux, E. Zh. B. Goursat, J. S. Hadamard, I. G. Petrovsky, S. L. Sobolev, S. V. Godunov, and many others. A feature of hyperbolic equations is the existence of non-differentiable solutions, which is associated with shock waves, which are described by such equations. This makes it often impossible to use standard analytical methods for solving boundary value problems (BVPs) using the smoothness properties of solutions. Numerical methods based on the difference approximation of equations, which also require the corresponding smoothness of solutions, often become ineffective.

One of the effective methods developed for solving BVPs of mathematical physics is the method of boundary integral equations (BIEM). BIEM is a direct development of potential theory. Through the work of outstanding scientists Gunter N. M., A. V. Bitsadze, N. P. Vekua, N. I. Muskhelishvili, S. G. Mikhlin, V. D. Kupradze, and others [1–7] were created a harmonious theory of singular integral equations for solving BVPs for elliptic equations and systems of mathematical physics. Currently, BIEM is one of the most intensively developed in the world of numerical and analytical methods for solving various problems of mathematical physics; however, it has received the greatest development in solving static, quasi-static, and stationary problems described by equations of elliptic and parabolic types.

The main advantage of the BIEM is that it reduces the size of the equations being solved by an order of magnitude, which is very important, especially for areas containing infinitely distant points. Instead of constructing solutions of differential equations in the domain of definition, the problem reduces to solving integral equations at the boundary of the region. This not only lowers the dimension of the problem, but also makes the procedure more stable for constructing a solution and its correct determination. Existing methods and programs for spline approximation of arbitrary contours and surfaces remove the problem of limiting the shape of the regions under consideration when using BIEM.

The method of boundary integral equations is based on the representation of a solution to a BVP through the boundary values of the desired function (or vector function) and certain combinations of its derivatives. In particular, solutions of the Laplace–Poisson equation inside a domain are expressed in terms of the boundary values of the desired function and its normal derivative using the well-known Green formula, which is the sum of the Newtonian potentials of a simple and double layer, whose densities are the indicated boundary functions. The study of the limiting properties of these potentials at the boundary of the region leads to resolving boundary integral equations relating the boundary values of the solution and its derivatives. The traditional theoretical method for constructing solutions of these BIEs in regions with arbitrary boundary geometry is the method of successive approximations, the convergence of which depends on the type of cores and the degree of smoothness of the boundary functions. This method in the theory of boundary value problems for elliptic systems is called the *direct method* for constructing the BIE.

The second approach to solving elliptic boundary value problems is given by potential theory. It is based on the representation of the solution in the form of potentials of a simple or double layer, which is a convolution of the fundamental solution of the corresponding differential equation with singular generalized functions—layers on the boundary of a region with unknown densities. Proceeding from the limiting properties of potentials, the BIEs are constructed to determine the unknown density. Both of these methods are very widespread for solving static problems and problems of stationary oscillations of the mechanics of an elastic body [4, 5, 7], problems of stationary wave diffraction in acoustics [8–10], electrodynamics [11–16], and other branches of mathematical physics.

We also note the intensive development in the last three decades of numerical methods for solving boundary integral equations, which are successfully used to solve various applied problems [17–19].

To solve nonstationary boundary value problems for hyperbolic equations based on these methods, the Laplace or Fourier transform in time and others is usually used, which reduce hyperbolic equations to elliptic-type equations in transformant space, parameterized in Laplace or Fourier variables. They are solved on the basis of a direct or indirect BIE method. To restore the original solutions, various numerical procedures of the inverse Fourier, Laplace, and other transformations are used. The class of such problems of mathematical physics began to be considered as early as the seventies of the last century, which was associated with the advent of computer technology, but so far the number of works in this direction is very limited.

As known, the inverse transformation procedures are unstable, which is typical for solutions of Fredholm equations of the first kind, and require regularization of the corresponding equations to obtain reliable results. Therefore, the problem of developing constructive methods for solving boundary value problems for hyperbolic equations and systems of equations of mathematical physics to study various wave processes in solids and continuous media remains relevant.

Researches in the field of boundary value problems for hyperbolic systems of equations and systems of mixed type are currently being carried out in all the leading mathematical schools of the world. It suffices to say that regularly (since 1985) international schools (once a year) and conferences (once every 2 years) are held: Hyperbolic equations: theory, numeric, and applications. The most developed theory of hyperbolic equations for three classical hyperbolic boundary value problems: the Cauchy problem, the Goursat problem, and the Darboux problem. These problems for the classical d'Alembert equation are included in all textbooks on mathematical physics (see [6, 20, 21] and many others). A further complication of these problems is associated with an increase in the dimension of spaces, a complication of the form of equations, and systems with the introduction of variable coefficients and new boundary conditions. A great contribution to this direction of research was also made by Kazakh mathematicians Kalmenov T. S., Temirbolat S. E., Aldashev S. A., Dzhumabaev D. S., Asanova A. T. [22–27], and others.

To solve nonstationary dynamic problems in continuous media described by systems of equations of hyperbolic type, various numerical methods have also been developed to date: the finite-difference method of spatial characteristics in the works of Clifton R. D. [28], Reker V. V. [29], and others, the method of disintegration of the gap in the works of Godunov S. V. [30] and others, fractional step method of Yanenko N. N. [31], the method of behavioral characteristics in the works of Tarabrin G. T., Baiteliev T. B., Mardonov B. M. [32, 33], and other numerical methods. However, the use of these numerical methods for solving hyperbolic problems is still limited by the requirement of differentiability of solutions, which requires the introduction of artificial smoothing of solutions in the field of shock wave fronts, the appearance of large gradients, and the associated instability of the procedure for constructing a solutions describing processes accompanied by shock waves which are characteristic for physical processes. In addition, great difficulties arise when using these methods in areas with arbitrary noncanonical boundaries.

However, BIEM allows to explore such processes. In works [34–39], a BIE method was developed for solving boundary value problems of the dynamics of elastic and electromagnetic media in regions with an arbitrary shape of boundaries. Here, the Generalized Functions Method (GFM) is used to construct dynamic analogues of Green's formulas and singular BIEs, which was first proposed by L. Alexeyeva in the work [34] for solving initial-boundary value problems for the equations of dynamics of elastic media and the wave equation [37, 38].

When constructing boundary integral equations for solving boundary value problems, fundamental solutions of equations are used. A characteristic feature of fundamental solutions of hyperbolic equations is that they belong to the class of singular generalized functions, which leads to the need to use the apparatus of the theory of generalized functions to develop the method of boundary integral equations based on it to solve nonstationary boundary value problems for a wide class of hyperbolic equations.

The solution of nonstationary dynamic problems on the basis of the BIE method requires the introduction of the concept of a solution in the space of generalized functions (generalized solution), which is associated with a feature of the fundamental solutions of wave equations that belong to the class of generalized functions. The main principles of this method, using the example of the classical multidimensional d'Alembert equation, were described in [40, 41], where dynamic analogs of Green's formulas and singular boundary integral equations were constructed in spaces of dimensions 1, 2, 3. In [38], for the regularization of the kernels of boundary integral equations, a regularization method on wave fronts was proposed, where they have strong non-integrable features.

Initially, this method was developed to solve the problems of wave diffraction in elastic and thermoelastic media. Later this method was used for regularization of BIEs in the space of generalized functions with the introduction of antiderivative of fundamental solutions, which allows one to construct solutions of initial-boundary value problems for systems of hyperbolic equations in spaces of arbitrary dimension [42–47].

In this monograph, the reader will be introduced to GFM and BIEM for classical hyperbolic equations and systems of equations of mathematical physics.

Chapter 1 is introductory to mastering the basic concepts and methods of the theory of generalized functions, which are further used to solve Boundary Value Problems (BVPs). In Chaps. 2–5, the theory of BVPs is constructed for classical field theory equations: the d'Alembert, Schrodinger, Klein–Gordon–Fock equations in spaces of dimensions $N = 1, 2, 3$, typical for problems of mathematical physics. Nonstationary processes and stationary ones with a constant frequency of oscillations are considered.

We also consider transport BVPs for the d'Alembert equation in cylindrical regions, which arise when the boundary functions move in the direction of the cylinder axis with a constant velocity v, subsonic or supersonic, depending on the ratio of v to the wave propagation velocity in the medium c: $v/c = M$. The type of equation in the moving coordinate system depends on the Mach number M: elliptic at subsonic speeds ($M < 1$) and hyperbolic at supersonic speeds ($M > 1$), which significantly affects the construction of the solution of the boundary value problem, one of which is elliptic and the second is hyperbolic. The solution of these problems on the basis of the GFM shows the universality of this method for solving elliptic and hyperbolic equations and systems.

The last three Chaps. 6–8 show how to use the generalized function method to solve systems of partial differential equations using Maxwell system of equations as an example. These classical equations of electrodynamics describe the relationship between electromagnetic field intensities, electric currents, and charges and consists of two vector equations for currents and two scalar equations for charges. For given currents, vector equations make it possible to determine the vectors of the electric and magnetic strength of the EM field. Once they are determined, scalar equations determine the charge through the divergence of these vectors [48, 49].

The traditional way of constructing solutions to Maxwell equations is associated with the transition to separate wave equations for the scalar and vector potentials of the electromagnetic field, the class of particular solutions of which is well studied. On their basis, a wide class of boundary value problems of electromagnetic dynamics has been solved in areas limited, as a rule, by coordinate surfaces in coordinate systems, where it is possible to use methods of complete and incomplete separation of variables. With an arbitrary shape of the boundary of electromagnetic waveguides, the transition to the potential form of representing solutions does not simplify, but even complicates the problem, since the order of the derivatives of the sought functions in the equation and boundary conditions is higher than in the original system. An effective method for solving such problems is the method of boundary integral equations.

The central point of BIE method is the construction of fundamental solutions of Maxwell equations, among which the Green tensor is of particular importance, since it and other tensors constructed on its basis are convenient to use when constructing the boundary integral equations as kernels of the corresponding integral operators. In Chap. 6, using a unified methodology based on the generalized Fourier transform, fundamental and generelized solutions to Maxwell equations are constructed in the space of generalized vector functions under the action of various currents and charges, nonstationary and stationary (harmonic in time). Periodic and non-periodic cases in the spatial variable that are important for applications are considered, for each of which the form of new Green tensors for an infinite medium is presented in a form convenient for calculations.

In Chaps. 7 and 8, solutions to stationary and nonstationary boundary value problems for Maxwell's equations in a limited domain are constructed, first in the space of generalized functions, then integral representations of magnetic and electric intensity vectors through their boundary values and, finally, singular boundary integral equations to determine these boundary values. Note that the apparatus of generalized functions allows one to naturally introduce representations for surface currents, charges, surface divergence, etc., the definition of which in classical electrodynamics does not satisfy the requirements of mathematical rigor. The use of methods of the theory of generalized functions significantly simplifies the procedure for constructing integral representations of solutions to boundary value problems, and allows one to reduce the requirements for the smoothness of functions, construct and study electromagnetic shock waves, and determine jumps on the fronts of shock EM waves.

It is well known that when constructing solutions to systems of equations of mathematical and theoretical physics, continuum mechanics often use scalar and vector potentials of vector fields that satisfy classical wave equations of different dimensions. The solution of boundary value problems for systems of interconnected differential equations can often be reduced to the solution of individual differential wave equations, which greatly simplifies the problem, since the class of solutions of the wave equation is well understood. Therefore, knowledge and the ability to solve boundary value problems for wave equations for determining the potentials of wave vector fields is of undoubted scientific interest and benefit.

The monograph can be recommended for study by master's and doctoral students of the Faculty of Mechanics and Mathematics of universities to master the methods of constructing solutions of equations of mathematical physics and boundary value problems. The authors hope that this monograph will help the reader in reading the works presented in the bibliography related to solving more complex boundary value problems for hyperbolic systems of equations describing the wave dynamics of elastic and electromagnetic media, and most importantly in their practical work related to the study of dynamic wave processes in solid bodies and continuous media.

References

1. Gunther, N.M.: Potential theory and its application to the main problems of mathematical physics—Moskow "Gostehizdat", 283 (1953)
2. Bitsadze, A.V.: Boundary value problems for second order elliptic equations—Moskow "Nauka", 203 (1966)
3. Vekua, N.P.: Systems of singular integral equations—Moskow "Nauka", 379 (1970)
4. Muskhelishvili, N.I.: Singular integral equations—Moskow "Nauka", 512 (1968)
5. Kupradze, V.D.: Potential methods in the theory of elasticity—Moskow "Nauka", 472 (1963)
6. Mikhlin, S.G.: Course of mathematical physics—Moskow "Nauka", 684 (1968)
7. Perlin, I.: Boundary integral equations in the theory of elasticity—Moskow "Nauka", 312 (1977)
8. Jones, D.S.: Integral equations for the exterior acoustic problem. Q. J. Mech. Appl. Math. **27**, 129–142 (1974)
9. Colton, D.: The inverse scattering problem for acoustic waves. Dundee Conference on Ordinary and Partial Differential Equations. Springer-Verlag Lecture Notes in Mathematics. **964**, 143–161 (1982)
10. Colton, D., Kress, R.: Iterative methods for solving the exterior Dirichlet problem for the Helmholtz equation with applications to the inverse scattering problem for low frequency acoustic waves. J. Math. Anal. Appl. **77**, 60–72 (1980)
11. Werner: On the exterior boundary value problem of perfect reflection for stationary electromagnetic waves. J. Math. Anal. Appl. **7**, 348–396 (1963)
12. Werner.: On an integral equation in electromagnetic diffraction theory. J. Math. Anal. Appl. **14**, 445–462 (1966)
13. Panich, O.I.: On the question of the solvability of external boundary value problems for the wave equation and Maxwell's equations. Advances in Mathematical Sciences. T. 20. 221–226 (1965)
14. Zakharov, E.V., Eremin, Y.: A. On a method for solving axisymmetric problems of diffraction of electromagnetic waves on bodies of revolution. Journal of Computational Mathematics and Mathematical Physics. T. 19, no. 5. 1344–1348 (1979)
15. Colton, D., Kress, R.: The impedance boundary value problem for the time harmonic Maxwell equations. Math. Meth. in the Appl. Sci. **3**, 475–487 (1981)
16. Colton, D., Kress, R.: Methods of integral equations in scattering theory—Moskow "Nauka". 311 (1987)
17. Brebbia, K., Walker, S.: Application of boundary elements in technology—Moskow "Mir". 248 (1982)
18. Brebbia, C., Telles, J., Braubel, L.: Boundary element methods—Moskow "Mir". 528 (1987)
19. Beskos, D.E.: Boundary element method in dynamic analyses. Appl. Mech. Rev. **40**(1), 1–23 (1987)
20. Courant, R., Hilbert, D.: Methods of mathematical physics—Moskow-Leningrad "Gostekhizdat". **1**, 184 (1951)
21. Vladimirov, V.S.: Equations of mathematical physics—Moskow "Nauka". 512 (1981)
22. Kalmenov, T.Sh.: Boundary value problems for linear partial differential equations of hyperbolic type-Shymkent "Gylym". 328 (1993)

23. Aldashev, S.A.: Boundary value problems for multidimensional hyperbolic and mixed equations—Almaty "Gylym". 170 (1994)
24. Dzhumabaev, D.S., Asanova, A.T.: Unique solvability of boundary value problems with data on characteristics for systems of hyperbolic equations. Computational mathematics and mathematics physics. **42**(11), 1673–1685 (2002)
25. Dzhumabaev, D.S., Asanova, A.T.: On the unique solvability of a nonlocal boundary value problem with data on intersecting lines for systems of hyperbolic equations. Differential equations. **45**(3). 373–381 (2009)
26. Dzhumabaev, D.S., Asanova, A.T.: Correct solvability of nonlocal boundary value problems for systems of hyperbolic equations. Differerential equations. **41**(3), 337–346 (2005)
27. Asanova, A.T.: On a nonlocal boundary value problem for systems of quasilinear hyperbolic equations. Reports of the Academy of Sciences. **411**(1), 7–11 (2006)
28. Clifton, R.J.: Difference method in plane problems of dynamic elasticity. Mechanics (collection of translations). 1, 103–122 (1968)
29. Reker: V.V.: Numerical solutions of three-dimensional problems of the dynamic theory of elasticity. Applied mechanics. Series E. 1, 121–129 (1970)
30. Godunov, S.K.: and others. Numerical solution of multidimensional problems of gas dynamics—Moskow "Nauka". 400 (1976)
31. Yanenko, N.N.: Methods of fractional steps for solving multidimensional problems of mathematical physics. Novosibirsk: Publishing house. NSU, 195 (1967)
32. Tarabrin, G.T.: Application of the bicharacteristics method for solving nonstationary problems in the dynamics of anisotropic massifs. Structural mechanics and design of structures. 4, 38–43 (1981)
33. Baiteliev, T.B., Mardonov, B.: Solution of a plane dynamic problem of the theory of elasticity for a strip by the method of spatial characteristics. In the book: Problematic issues of rock mechanics. Alma-Ata. 194–201 (1972)
34. Alexeyeva, L.A.: Analogues of the Kirchhoff-Somigliana formulas in plane problems of elastodynamics. Applied mathematics and mechanics. **55**(2), 298–308 (1991)
35. Alexeyeva, L.A.: Boundary Element Method of Boundary Value Problems of Elastodynamics by Stationary Running Loads. Int. J. Engineering Analysis with Boundary Element. 11, 37–44 (1998)
36. Alexeyeva, L.A., Sautbekov, S.S.: Method of generalized functions in stationary boundary value problems for Maxwell's equations//Journal of Computational Mathematics and Mathematical Physics. **40**(4), 611–622 (2000)
37. Alexeyeva, L.A.: Dynamic analogues of the Green and Gauss formulas for solutions of the wave equation in Differential equations. **31**(11), 1951–1953 (1995)
38. Alexeyeva, L.A.: Boundary integral equations of the initial-boundary value problem for the wave equation in Differential equations. **28**(8), 1451–1453 (1992)
39. Alexeyeva, L.A.: Generalized solutions of nonstationary boundary value problems for Maxwell's equations. Journal of Computational Mathematics and Mathematical Physics. **42**(1) 76–88 (2002)
40. Alexeyeva, L.A.: Method of generalized functions in nonstationary boundary value problems for the wave equation. Mathematical Journal. **6**1(19), 16–32 (2006)
41. Alexeyeva, L.A.: Boundary integral equations of nonstationary BVP for wave equations. Int. Congress of Mathematicians. Abstracts. Madrid, 436 (2006)
42. Dynamic analogues of the Somilyana formula for nonstationary dynamics of elastic media with an arbitrary degree of anisotropy. Applied Mathematics and Mechanics. **58**(2), 170–175 (1994)
43. Alexeyeva, L.A., Dildabayev, Sh.A., Zakiryanova, G.K., Zhanbyrbaev, A.B.: Boundary Integral Equations Method in two-and three-dimensional problems of elastodynamics. Computational mechanics. **18**, 147–157 (1996)
44. Alexeyeva, L.A., Zakiryanova, G.K.: Generalized Solutions of Boundary Value Problems of Dynamics of Anisotropic Elastic Media//Journal of the Mechanical Behavior of Materials. **16**(4–5), 259–267 (2005)

45. Alexeyeva, L.A., Kupesova, B.N.: Method of generalized functions in boundary value problems of coupled thermoelastodynamics. Applied Mathematics and Mechanics. **65**(2), 334–345 (2001)
46. Alexeyeva, L.A., Zakiryanova, G.K.: Green's matrix for strictly hyperbolic systems with second-order derivatives. Differential equations. **37**(4), 488–494 (2001)
47. Alexeyeva, L.A., Zakiryanova, G.K.: Generalized solutions of initial-boundary value problems for second-order hyperbolic systems. Journal of Computational Mathematics and Mathematical Physics. **51**(7), 1280–1293 (2011)
48. Tamm, I.E.: Fundamentals of the theory of electricity. M.: Nauka, (1989)
49. Feynman, R.P., Leighton, R.B, Sands, M.: The Feynman Lectures on Physics, v. 2: The Electromagnetic Field—Addison-Wesley, (1963)

Notations

BIE	Boundary Integral Equation
BVP	Boundary Value Problem
GFM	Generalized Functions Method
KGF	Klein–Gordon–Fock equations
MBIE	Method of Boundary Integral Equations
RGF	Regular Generalized Function
SGF	Singular Generalized Function
R^N	Euclidean space of N dimension
$C(.)$	Continuous function set on the set of variables in parentheses
$C^k(.)$	Continuous functions together with derivatives up to the k-th order
$L_k(.)$	Lebesgue space
$D(R^N)$	A space of finite infinitely differentiable functions
$D'(R^N)$	A space of generalized functions on $D(R^N)$
$S(R^N)$	Schwarz space of functions rapidly decreasing at infinity
$S'(R^N)$	A space of generalized functions of slow growth on $S(R^N)$

$x = (x_1, \ldots, x_N) \in R^N$

$\|x\| = \sqrt{x_1^2 + \ldots + x_N^2} \triangleq \sqrt{x_j x_j}$	A norm of the x in R^N
(x, y)	Scalar product
$[x, y]$	Vector product
$\frac{\partial u}{\partial t} = \dot{u} = \partial_t u$	A derivative in time
$f_{,i} = \partial_i f = \frac{\partial f}{\partial x_i}$	Partial derivative in x_i

$\nabla = \text{grad}$

$\Delta = \sum_{j=1}^{N} \frac{\partial^2}{\partial x_j^2}$ Laplace operator (*Laplacian*)

$\Delta = \sum_{j=1}^{N} \frac{\partial^2}{\partial x_j^2}, \quad n < N$

$\Box_c = \Delta - \frac{1}{c^2}\frac{\partial^2}{\partial t^2}$	Wave operator (*d'Alembertian*)
S	Boundary of S^- from the class of Lyapunov surfaces
S^-	Open simply connected set in R^N
$n(x) = \{n_1(x), ..., n_N(x)\}$	A normal to $S \subset R^N$, $\\|n\\| = 1$
$\hat{u}(x)$	Generalized function—continuous linear functional on $D(R^N)$or $S(R^N)$
$(\hat{u}, \varphi)$	A value of $\hat{u}$ on $\varphi \in D(R^N)$ or $\varphi \in S(R^N)$
$\alpha(x)\delta_S(x)$	Simple layer on S, singular generalized function
$\alpha(x)$	Simple layer density
$H(t)$	Step Heaviside function
$H_S^-(x)$	Characteristic function of a set S^-

Contents

Chapter 1
Basic Concepts of Generalized Functions Theory: Fundamental Solutions and Their Properties

The main idea of the method of generalized functions (GFM) in solving boundary value problems for differential equations is the transition from the original boundary value problem (BVP), in which a differential equation or a system of differential equations and boundary conditions are specified on the boundary of the region, the number of which depends on the order of the equations (for external boundary value problems are also given certain conditions at infinity) to differential equations in the space of generalized functions with the right-hand side containing singular generalized functions of the type simple, double, or more layers at the boundary of the domain of definition, the densities of which depend on the boundary values of the solution and its derivatives. Fundamental solutions of these equations that satisfy certain conditions of attenuation and radiation can be used to construct a solution to the original boundary value problem in the form of a convolution in the space of generalized functions with a singular right-hand side. The regular integral representation of this convolution gives a solution within the domain of definition and divergent integrals on its boundary. In addition, the study of the limiting properties of these representations when approaching the boundary of the region allows us to construct resolving singular boundary integral equations for determining unknown boundary functions.

In this chapter, we introduce the notation and define some concepts of the method of generalized functions, which we will use in the future when constructing solutions of the boundary value problems proposed to the reader. Here we use a lot of notation adopted by the founders of the theory of generalized functions [1–5], in particular, V. Vladimirov, who built the fundamental solutions of many equations of mathematical physics and wrote an excellent textbook on mathematical physics with a summary of the foundations of this theory and its applications [4]. In addition, the problem book on mathematical physics developed under his leadership [5] serves as an excellent reference book on mathematical physics, to which very authoritative publications refer [6].

L. Alexeyeva and A. Bayegizova, *Generalized Functions Method in Boundary Value Problems for Wave Equations*, Industrial and Applied Mathematics,
https://doi.org/10.1007/978-981-95-2802-8_1

1.1 Regular and Singular Generalized Functions

Let us denote $D'(R^N) = \{\hat{u}(x), x \in R^N\}$ is the space of generalized functions defined on the space of infinitely differentiable functions $D(R^N) = \{\varphi(x), x \in R^N, \varphi \in C^\infty(R^N)\}$ with finite support [5]. Any locally integrable function $f(x)$ defines a *regular generalized function* (RGF) $\hat{f}(x)$. It is a linear continuous functional of the form:

$$(\hat{f}, \varphi) = \int_{R^N} f(x)\varphi(x)\, dV(x),\ dV(x) = dx_1 dx_2 ... dx_N. \tag{1.1}$$

Generalized functions that cannot be represented in the form (1.1) are called *singular.*

Definition 1 A singular generalized function that defines a continuous linear functional of the form.

$$(\delta(x), \varphi(x)) \equiv \varphi(0)$$

is called δ-*function.* It has the following properties:

$$\delta(x) = \delta(-x), \quad x \in R^N, \tag{1.2}$$

$$f(x)\delta(x) = f(0)\delta(x), \tag{1.3}$$

$$\delta(\lambda x) = \frac{1}{|\lambda|^N}\delta(x), \tag{1.4}$$

$$\delta(x) = \delta(x_1) \cdot ... \cdot \delta(x_N). \tag{1.5}$$

Here there is a direct product of functions in the last formula.

When constructing solutions to boundary value problems, we use singular generalized functions $\alpha(x)\delta_S(x)$. It is *a simple layer* on the surface $S \subset R^N$, which defines the linear functional of the form

$$(\alpha(x)\delta_S(x), \varphi(x)) \overset{\Delta}{=} \int_S \alpha(x)\varphi(x) dS(x) \tag{1.6}$$

and a *double layer*

$$\frac{\partial}{\partial n}(\beta(x)\delta_S(x)) \triangleq \sum_{j=1}^{N} \partial_j\big(n_j(x)\beta(x)\delta_S(x)\big),$$

$$\left(\sum_{j=1}^{N} \partial_j\big(n_j(x)\beta(x)\delta_S(x)\big), \varphi(x)\right) \triangleq -\int_S \beta(x)\frac{\partial \varphi}{\partial n(x)} dS(x), \tag{1.7}$$

where $\alpha(x)$, $\beta(x)$ are the *densities of simple and double layers.* They are integrable functions on S, and $n(x)$ is a single normal to the surface S.

The formula

$$\big(\partial_i \hat{f}, \varphi\big) = -\left(\hat{f}, \frac{\partial \varphi}{\partial x_i}\right), \quad i = 1, ..., N, t \tag{1.8}$$

defines the *derivative* of a generalized function, which we call the *generalized derivative,* in contrast to the classical derivative. Let us denote partial derivative $D^\alpha = \frac{\partial^{|\alpha|}}{\partial x^{\alpha_1} \partial x^{\alpha_2} ... \partial x^{\alpha_N}}$, where the multiindex $\alpha = (\alpha_1, \alpha_2, ..., \alpha_N)$, α_j are positive numbers, and $|\alpha| = \sum_{k=1}^{N} \alpha_k$. From (1.8) follows the definition of a generalized derivative of any order:

$$\big(D^\alpha \hat{f}, \varphi\big) = (-1)^{|\alpha|}\big(\hat{f}, D^\alpha \varphi\big). \tag{1.9}$$

If $f(x)$ is regular and differentiable everywhere except for a surface S on which it suffers a discontinuity of the first kind, then its generalized derivative is equal to

$$\partial_j \hat{f}(x) \triangleq \frac{\partial \hat{f}(x)}{\partial x_j} = \frac{\partial f(x)}{\partial x_j} + n_j(x)\big[f(x)\big]_F \delta_S(x), \tag{1.10}$$

where the first term is the classical derivative, $[f]_S$ is a jump on S:

$$[f]_S = \lim_{\varepsilon \to +0} \{f(x + \varepsilon n(x)) - f(x - \varepsilon n(x))\}, \quad x \in S, \tag{1.11}$$

where $n(x) = \sum_{j=1}^{n} n_j(x) e_j$ is a unit normal vector n to S.

In formula (1.10), the first term on the right is the classical derivative. Here and below, we mark the function from above with a "hat" (^), when we consider it as a generalized one.

1.2 The Characteristic Function of a Set and Its Properties

Let us denote $S^- \subset R^N \quad (\dim\ S^- = N)$ as the open set in Euclidean space R^N; S is the boundary of this set ($\dim S = N - 1$) from the class of Lyapunov surfaces: $\exists\beta : 0 < \beta \le 1$, so that for $\forall x \in S,\ \forall y \in S$ has an assessment:

$$\|n(x) - n(y)\| \le const\|x - y\|^{\beta}, \tag{1.12}$$

where $n(x)$ is the outward unit normal to S at a point x.

Let us denote by $H_S^-(x)$ the characteristic function of the set S^-:

$$H_S^-(x) = \begin{cases} 1,\ x \in S^-, \\ 0,\ x \in S^+, \\ \frac{1}{2},\ x \in S, \end{cases} \tag{1.13}$$

where $S^+ = R^N - \left(S + S^-\right)$. RGF $H_S^-(x) \in D'\left(R^N\right)$ defines a functional of the form:

$$\left(H_S^-(x), \varphi(x)\right) = \int\limits_{S^-} \varphi(x)\,dV(x).$$

Its generalized derivatives are defined by the following Lemma.

Lemma 2.1 *Generalized derivative of $H_S^-(x)$ is equal to*

$$\frac{\partial H_S^-(x)}{\partial x_j} = -n_j(x)\delta_S(x), \tag{1.14}$$

where $n(x) = \left\{n_j(x),\ j = 1, \dots, N\right\}$ is an external normal $n(x)$ to S^-in a point x at S, $\|n\| = 1$. Singular generalized function $n_j(x)\delta_S(x)$ is a simple layer on S.

Proof Using the definition of a simple layer:

$$\left(n_j\delta_S(x), \varphi(x)\right) \triangleq \int\limits_S n_j(x)\varphi(x)dS(x)$$

and the Ostrogradsky–Gauss formula, we calculate

$$\left(\frac{\partial H_S^-(x)}{\partial x_j}, \varphi(x)\right) \triangleq -\left(H_S^-(x), \frac{\partial \varphi(x)}{\partial x_j}\right) = -\int\limits_{R^N} H_S^-(x)\frac{\partial \varphi(x)}{\partial x_j}dV(x)$$

$$= -\int\limits_{S^-} \frac{\partial \varphi(x)}{\partial x_j} dV(x) =$$

$$= -\int\limits_{S} \varphi(x) n_j(x) dS(x) = \big(-n_j(x)\delta_S(x), \varphi(x)\big).$$

The Lemma is proven.

Let $u(x) \in L_1(S^- + S)$ and supp $u(x) \in (S^- + S)$. We introduce a generalized function $\hat{u}(x)$, having defined by zero outside the domain of definition, more precisely:

$$\hat{u}(x) = u(x) H_S^-(x) \triangleq \begin{cases} u(x), & x \in S^-, \\ 0, & x \in S^+, \\ \frac{1}{2} u(x), & x \in S. \end{cases} \tag{1.15}$$

Let us define the generalized derivatives of $\hat{u}(x)$.

Lemma 2.2 *Generalized derivatives of $\hat{u}$ are equal to*

$$\frac{\partial \hat{u}}{\partial x_j} = \frac{\partial u}{\partial x_j} H_S^-(x) - u_S(x) n_j(x) \delta_S(x), \tag{1.16}$$

$$\frac{\partial^2 \hat{u}}{\partial x_j \partial x_i} = \frac{\partial^2 u}{\partial x_j \partial x_i} H_S^-(x) - \frac{\partial u}{\partial x_j} n_i(x) \delta_S(x) - \partial_i \big(u_S(x) n_j(x) \delta_S(x)\big), \tag{1.17}$$

where the first term on the right is the classic derivative, and $u_S(x)$ is the value $u(x)$ on S.

Proof Using the definition of the derivative of a generalized function, we have

$$\left(\frac{\partial \hat{u}}{\partial x_j}, \varphi(x)\right) \triangleq -\left(\hat{u}, \frac{\partial \varphi(x)}{\partial x_j}\right) = -\left(u(x) H_S^-(x), \frac{\partial \varphi(x)}{\partial x_j}\right)$$

$$= -\int\limits_{S^-} u(x) \frac{\partial \varphi(x)}{\partial x_j} dV(x) =$$

$$= -\int\limits_{S^-} \left(\frac{\partial \{u(x)\varphi(x)\}}{\partial x_j} - \varphi(x) \frac{\partial u(x)}{\partial x_j}\right) dV(x) =$$

$$= -\int\limits_{S} u(x)\varphi(x) n_j(x) dS(x) + \int\limits_{S^-} \varphi(x) \frac{\partial u(x)}{\partial x_j} dV(x) =$$

$$=\big(-u(x)n_j(x)\delta_S(x),\varphi(x)\big)+\left(\frac{\partial u(x)}{\partial x_j}H_S^-(x),\varphi(x)\right).$$

The first formula of the Lemma is proven. The second formula is a consequence of the first. The Lemma is proven.

Corollary to Lemma 2.2 is the action of Laplace operator on $\hat{u}$ in the GF space.

Lemma 2.3 *If* $u(x)\in C^2\big(S^-+S\big)$, *then*

$$\Delta\hat{u}=H_S^-(x)\Delta u-\frac{\partial u}{\partial n}\delta_S(x)-\partial_j\big(u_S(x)n_j(x)\delta_S(x)\big). \tag{1.18}$$

Hereinafter, the sign of the sum over repeated indices is omitted, as in a tensor convolution.

This property of the Laplace operator is convenient to use in the construction of generalized solutions of differential equations.

1.3 Fundamental Solutions of Differential Equations and Their Properties

We denote linear differential operator with constant coefficients

$$L(\partial_x)=\sum a_\gamma D^\gamma.$$

Let us consider the differential equation on D':

$$L(\partial_x)\hat{u}(x)=\hat{f}(x). \tag{1.19}$$

Among the generalized solutions of differential equations, fundamental solutions (FS) are of particular importance by construction solutions of differential equations. They describe the state of the medium under the action of sources concentrated at a point $x=0$.

Definition The generalized solution of the equation

$$L(\partial_x)\hat{U}(x)=\delta(x) \tag{1.20}$$

is called *the fundamental solution* of this equation.

Fundamental solutions are determined ambiguously, up to the solution of a homogeneous equation. They play a fundamental role in constructing solutions of boundary value problems. For their construction, the generalized Fourier transform (FT) is used, which is determined for the generalized functions of slow growth by the relation:

$$(F[\hat{f}], \varphi) \triangleq (\hat{f}, F[\varphi]), \quad \forall\varphi \in S'(R^N),$$

where

$$F_x[\varphi](\xi) = \varphi^*(\xi) = \int\limits_{R^N} \varphi(x)e^{i(\xi,x)}dV(x).$$

Here $S'(R^N)$ is the space of infinitely differentiable functions that rapidly decrease at $\|x\| \to \infty$: $\varphi = o\left(\frac{1}{(1+\|x\|)^k}\right), \quad \forall k > 0$. For regular functions for which

$$f = o\left((1 + \|x\|)^{-1}\right) \text{ at } \|x\| \to \infty,$$

the generalized Fourier transform coincides with the classical:

$$F[\hat{f}(x)] = f^*(\xi) = \int\limits_{R^N} f(x)e^{i(\xi,x)}dV(x),$$

$$\hat{f}(x) = \frac{1}{(2\pi)^N}\int\limits_{R^N} f^*(\xi)e^{-i(\xi,x)}dV(\xi).$$

If the support of a generalized function is bounded then

$$F_x[\hat{f}(x)] = (\hat{f}(x), \exp(i(\xi, x\pi)).$$

In particular,

$$F_x[\alpha(x)\delta_S(x)] = \int\limits_S \alpha(x)\exp(i(\xi, x)dS(x)). \tag{1.21}$$

Fourier transform property of the derivative

$$F_x\left[\partial_j\hat{f}(x)\right] = -i\xi_j F_x\left[\hat{f}(x)\right] = -i\xi_j f^*(\xi)$$

allows to build Fourier transforms of solutions of differential equations. So, since

$$F_x[\delta(x)] = 1,$$

from (1.20) this follows $F_x\left[\hat{U}(x)\right] = (L(-i\xi))^{-1}$.

To restore the original, the inverse Fourier transform is used:

$$F_{\xi}^{-1}\left[f^{*}(\xi)\right] = (2\pi)^{-N} F_{\xi}\left[f^{*}(-\xi)\right] = \hat{f}(x).$$

For basic functions, it has a classic look:

$$F_{\xi}^{-1}\left[\varphi^{*}(\xi)\right] = (2\pi)^{-N} \int_{R^N} \varphi^{*}(\xi) e^{i(\xi,x)} dV(\xi) = \varphi(x).$$

It is not possible to use this formula for the determination of $U(x)$, since the polynomial $L(-i\xi)$ usually has real roots on the surface in R^N, on which $L(-i\xi) = 0$. Therefore, the original function is ambiguous. This is due to the fact that the corresponding homogeneous differential equation has nonzero solutions. To restore the original, it is necessary to regularize the resulting Fourier transform, which, generally speaking, does not belong to the class of generalized functions. For this, additional properties of fundamental solutions should be used, in particular, the required asymptotic behavior at infinity.

The fundamental solution has a remarkable property, namely a generalized solution of Eq. (1.19) can be represented in the form:

$$\hat{u} = \hat{u}_0 + \hat{U} * \hat{f}, \tag{1.22}$$

where $\hat{u}_0$ is the generalized solution of the homogeneous equation ($f = 0$), and the second term is the convolution of the generalized functions. For regular functions, convolution can be represented in integral form:

$$\hat{U} * \hat{f} = \int_{R^N} U(x-y) f(y) dV(y). \tag{1.23}$$

For singular convolutions in the calculation, use the convolution definition (see [5]):

$$\left(\hat{U} * \hat{f}, \varphi\right) = \lim_{m\to\infty} (U(x) f(y), h_m(x,y)\varphi(x+y)), \tag{1.24}$$

where $h_m(x,y) \to 1$ at $m \to \infty$, $h_m(x,y) \in D(R^{2N})$.

Note that convolution does not always exist, which imposes certain restrictions on the right-hand side of (1.19) depending on the choice of the fundamental solution. Convolution always exists if one of the functions has a destination medium. It has a very useful property:

$$\partial_j\left(\hat{U} * \hat{f}\right) = \partial_j \hat{U} * \hat{f} = \hat{U} * \partial_j \hat{f}, \tag{1.25}$$

which we will use when differentiating convolution, choosing the most convenient form for calculation.

Generalized solutions of differential equations with a regular right-hand side are often regular and even differentiable functions almost everywhere. In such cases, they can also be considered as classical solutions of these equations based on the Du Bois-Reymond lemma, according to which, if two regular generalized functions are equal, then they are almost everywhere equal (p.95 [5]). This lemma allows to pass from generalized solutions to classical ones if they are regular.

1.4 Representation of Characteristic Function of a Set: Generalized Analogue of Gauss Formula

If we take convolution of Eq. (1.20) with the characteristic function of the set $H_S^-(x)$ and to use differentiation property of the convolution (1.25), then we obtain a representation of $H_S^-(x)$ through the fundamental solution of this differential equation.

Lemma 4.1 (Generalized Gauss formula): *The characteristic function of the domain S^- is representable as a convolution:*

$$H_S^-(x) = \hat{U}(x) * L(\partial_x)H_S^-(x),$$

where $\hat{U}(x)$ is fundamental solution of the operator $L(\partial_x)$.

The integral representation of the formula depends on the form of the differential operator and its fundamental solution. Presenting it for boundary points requires special proof, since this requires a study of the asymptotic properties of $\hat{U}(x)$ at $x \to 0$. In particular, for Laplace operator $L(\partial_x) = \Delta = \sum_{k=1}^{N} \frac{\partial^2}{\partial x_k^2}$, this formula gives the famous *Gauss formula:*

$$H_S^-(x) = \int_S \frac{\partial U(x-y)}{\partial n(y)} dS(y). \tag{1.26}$$

Indeed, using the convolution differentiation property (1.25), we obtain

$$H_S^-(x) = \Delta\big(U(x) * H_S^-(x)\big) = -\sum_{m=1}^{N} \partial_m U(x) * n_m \delta_S(x) =$$
$$= -\int_S n_m(y) \frac{\partial U(x-y)}{\partial x_m} dS(y) = \int_S \frac{\partial U(x-y)}{\partial n(y)} dS(y).$$

Here $U(x)$ is the fundamental solution of the Laplace–Poisson equation:

$$\Delta U(x) = \delta(x).$$

That has the following form:

$$U(x) = \frac{1}{2\pi} \ln \frac{1}{\|x\|}, \quad x \in R^2;$$
$$U(x) = -\frac{1}{\sigma_N \|x\|^{N-2}}, \quad x \in R^N, \ N \geq 3, \tag{1.27}$$

where σ_N is an area of a sphere of radius 1 in R^N: $\sigma_N = \frac{2\pi^{N/2}}{\Gamma(N/2)}$, $\Gamma(z) = \int\limits_0^\infty t^{z-1}e^{-t}dt$ is gamma function.

For boundary points $x \in S$, the integral in (1.26) is taken in the sense of the principal value and is equal to ½, i.e., it gives the value of the characteristic function of the set on the smooth boundary S in the sense of its definition (1.13).

1.5 Stationary Vibrations: Analogue of Gauss Formula for Fundamental Solution of Helmholtz Equation

A special class in the problems of mathematical physics is solutions that describe stationary harmonic oscillations of the medium. They can be represented in the form: $u(x, t) = u^*(x)e^{-i\omega t}$, where ω is the oscillation frequency, $u^*(x)$ is a complex amplitude, the modulus of which is equal to the amplitude of the oscillations, and the argument describes the phase shift of the oscillation points of the medium. In this case, if $u^*(x)e^{-i\omega t}$ is the solution of the d'Alembert equation with the right side $G^*(x)e^{-i\omega t}$, then $u^*(x)$ is a solution to the Helmholtz equation:

$$\Delta u^* + k^2 u^* = G^*(x), \quad k = \omega/c. \tag{1.28}$$

The Helmholtz equation in oscillation theory plays the same role as the Laplace equation in static problems of mathematical physics. The class of particular solutions of the homogeneous Helmholtz equation, which is described by cylindrical and spherical Bessel functions, has been widely studied [7].

Equation (1.28) has fundamental solutions of the form:

$$\hat{U}_1 = -\frac{i}{4} H_0^1(k\|x\|), \quad \hat{U}_2 = -\frac{i}{4} H_0^2(k\|x\|), \quad N = 2; \tag{1.29}$$

$$\hat{U}_1 = -\frac{e^{ik\|x\|}}{4\pi\|x\|}, \quad \hat{U}_2 = -\frac{e^{-ik\|x\|}}{4\pi\|x\|}, \quad N = 3. \tag{1.30}$$

Here $H_0^1(k\|x\|)$, $H_0^2(k\|x\|)$ are cylindrical Hankel functions of the first and second kind [7].

The choice of the fundamental solution in constructing the solutions of the inhomogeneous Eq. (1.28) depends on the radiation conditions at infinity. In particular, when constructing solutions describing the waves diverging and decaying at infinity, the Sommerfeld radiation conditions are used [5, 8, 9]. They have the following form: at $r = \|x\| \to \infty$

$$u(x) = O\left(r^{-1/2}\right),\ \tfrac{\partial u}{\partial r} - iku = o\left(r^{-1/2}\right) \quad \text{for } N = 2, \tag{1.31}$$

$$u(x) = O\left(r^{-1}\right)\tfrac{\partial u}{\partial r} - iku = o\left(r^{-1}\right) \quad \text{for } N = 3. \tag{1.32}$$

If $u(x,t) = u(x)e^{-i\omega t}$, then Sommerfeld conditions are satisfied for $\hat{U}_1$. The corresponding convolution $\hat{U}_1(x) * G^*(x)$ will satisfy them, provided that it exists.

Let L^- be an arbitrary closed domain in R^N bounded by a closed Lyapunov contour (surface) L, and $n(x)$ is unit external normal to L, $L^+ = R^N - \left(L + L^-\right)$.

Lemma 5.1 *The stationary analogue of Gauss formula for Helmholtz equation has the form:*

$$\text{V.P.} \int_L \tfrac{\partial U(y-x)}{\partial n(y)} dS(y) + k^2 \int_{L^-} U(x-y)dV(y) = H_L^-(x),$$

where L^- is arbitrary region in R^N, bounded by a closed Lyapunov contour (surface) L, $n(x)$ is the unit external normal to L. For $x \in S$ the integral is taken in the sense of Value Principle.

Proof: Let us take convolution Eq. (1.28) at $\hat{G} = \delta(x)$ with $H_L^-(x)$. Using the convolution differentiation property and Lemma 2.1, we obtain

$$-\sum_{j=1}^{N} \frac{\partial U}{\partial x_j} * n_j \delta_L(x) + k^2 U * H_L^-(x) = H_L^-(x). \tag{1.33}$$

The integral representation of this equality has the form:

$$\int_L \tfrac{\partial U(y-x)}{\partial n(y)} dS(x) + k^2 \int_{L^-} U(x-y)dV(y) = H_L^-(x). \tag{1.34}$$

For $x \notin S$ the integrand functions are continuous; since $y \neq x$ for $\forall y \in S$, consequently $r = \|y - x\| \neq 0$. Therefore, all integrals exist.

For $x \in S$ integrands have a singularity when $y = x$ (endless at this point). Moreover, in the first integral there is a weak integrable singularity, by virtue of

(1.33, 1.34), and in the second surface integral $\frac{\partial U(y,x)}{\partial n(y)}$ has a strong non-integrable singularity when $y = x$

$$\frac{\partial U(y-x)}{\partial n(y)} = \frac{\partial U}{\partial r}\frac{\partial r}{\partial n(y)} = \frac{\partial U}{\partial r}\left(n(y), \frac{y-x}{r}\right),$$

$$\frac{\partial U}{\partial r} = -\frac{ik}{4}H_1^1(kr), \frac{\partial U}{\partial r} = -\frac{ik}{4r} \text{for } r \to 0, \quad N = 2, \tag{1.35}$$

$$\frac{\partial U}{\partial r} = -\frac{e^{ikr}}{4\pi r}\left(\frac{1}{r} - ik\right), \quad \frac{\partial U}{\partial r} \sim -\frac{1}{4\pi r^2} \quad \text{for } r \to 0, \quad N = 3. \tag{1.36}$$

We will show that the integral also exists in this case, if it is calculated in the sense of the principal value.

If L^- is a ball of radius ε, a $x^*$is the center of the ball, it follows from the formula

$$\int\limits_{r=\delta} \frac{\partial U(y, x^*)}{\partial n(y)} dS(y) + k^2 \int\limits_{r<\delta} U(y, x^*) dV(y) = 1, r = \left\| y - x^* \right\|.$$

Passing to the limit on $\varepsilon \to 0$, we obtain

$$\lim_{\varepsilon\to 0} \int\limits_{r=\varepsilon} \frac{\partial U(y, x^*)}{\partial n(y)} dS(y) + k^2 \lim_{\varepsilon\to 0} \int\limits_{r<\varepsilon} U(y, x^*) dV(y) =$$
$$= \lim_{\varepsilon\to 0} \int\limits_{r=\varepsilon} \frac{\partial U(y, x^*)}{\partial n(y)} dS(y) = 1.$$

Since

$$\frac{1}{2\pi} \lim_{\varepsilon\to 0} \int\limits_{r<\varepsilon} r \ln r dr d\vartheta = 0, \quad N = 2;$$

$$\frac{1}{4\pi} \lim_{\varepsilon\to 0} \int\limits_{r<\varepsilon} r^{-1} r^2 \sin\vartheta dr d\vartheta d\varphi = 0, \quad N = 3,$$

so $\lim\limits_{\varepsilon\to 0} \left| \int\limits_{r<\varepsilon} U(y - x^*) dV(y) \right| < \lim\limits_{\varepsilon\to 0} \int\limits_{r<\varepsilon} |U(y - x^*)| dV(y) = 0.$

Consequently,

$$\lim_{\varepsilon\to 0} \int\limits_{r=\varepsilon} \frac{\partial U(y - x^*)}{\partial n(y)} dS(y) = 1.$$

Due to the properties of symmetry and equality

$$n(x + e\varepsilon) = -n(x - e\varepsilon), \|e\| = 1,$$

for any hemisphere (semicircle) $\Gamma_\varepsilon(x^*)$ radius ε with center x^*

$$\lim_{\varepsilon\to 0} \int\limits_{y\in\Gamma_\varepsilon(x^*)} \frac{\partial U(y, x^*)}{\partial n(y)} dS(y) = 0,5. \tag{1.37}$$

We use this equality to prove the lemma for boundary points.

We introduce the following notation: $r_{,j} = \frac{\partial r}{\partial y_j} = \frac{y_j - x_j}{r}$, $\frac{\partial r}{\partial n(y)} = \sum_{j=1}^{N} r_{,j} n_j(y)$, $\Gamma_\varepsilon^\pm(x) = \{y \in L^\pm : r = \varepsilon\}$, $Ш_\varepsilon^\pm = L^\pm \cap Ш_\varepsilon$, $L_\varepsilon^\pm(x) = L^\pm - Ш_\varepsilon^\pm(x)$, $Ш_\varepsilon(x) = \{y : r < \varepsilon\}$, $O_\varepsilon(x) = \{y \in L : r \le \varepsilon\}$, $L_\varepsilon(x) = L - O_\varepsilon(x)$.

Let us transform the contour L in the vicinity of the boundary point $x \in L$, by passing it along the semicircle in L^-. We write a dynamic analog of the Green formula for the contour $L_\varepsilon + \Gamma_\varepsilon^+(x)$, for which the point x is internal:

$$\int\limits_{O_\varepsilon(x)} \frac{\partial U(y-x^*)}{\partial n(y)} dS(y) + k^2 \int\limits_{L^-} U(y-x^*) dV(y) +$$

$$+ \int\limits_{\Gamma_\varepsilon^+(x)} \frac{\partial U(y-x^*)}{\partial n(y)} dS(y) + k^2 \int\limits_{Ш_\varepsilon^+(x)} U(y-x^*) dV(y) = 1$$

Passing to the limit by $\varepsilon \to 0$ in this formula, taking into account the definition of the integral in the sense of the principal value, we obtain

$$\text{V.P.} \int\limits_{L} \frac{\partial U(y, x)}{\partial n(y)} dS(y) + k^2 \int\limits_{L^-} U(y, x) dV(y) + 0,5 = 1.$$

Since for the boundary points $H_L^-(x) = 1/2$, from here we obtain the formula of the lemma for boundary points.

From the formula of this lemma, in view of the asymptotic behavior of $U(x)$ at zero, we obtain a corollary useful for further arguments:

$$\lim_{\varepsilon\to +0} \int\limits_{\|y-x\|=\varepsilon} \frac{\partial U(y,x)}{\partial n(y)} dS(y) = 1\,.$$

This property should be used in studying the limiting properties of the integral representations of convolutions on the boundary of the solution domain of boundary value problems for solutions of the stationary Schrodinger–Gordon–Fock equation.

The Dirichlet and Neumann boundary value problems for the Helmholtz equation belong to the class of well-studied ones [9–11].

1.6 Fundamental Solutions of D'Alembert Equation: Simple Layer on a Light Cone

Let us consider the fundamental solutions of d'Alembert equation—classical wave equation:

$$\Box_c \hat{U} \equiv \Delta \hat{U} - c^{-2}\frac{\partial^2 \hat{U}}{\partial t^2} = \delta(x)\delta(t), \quad x \in R^N, \ t \in R^1. \tag{1.38}$$

Their form substantially depends on the dimension of space N.

In one-dimensional and two-dimensional spaces, fundamental solutions are regular GFs; their support is the interior of the cone: $\|x\| < ct$:

$$\hat{U}(x, t) = -\frac{c}{2}H(ct - |x|), \quad N = 1, \tag{1.39}$$

where $H(t)$ is Heaviside step function:

$$\hat{U}(x, t) = -\frac{H(ct - r)}{2\pi\sqrt{t^2 - (r/c)^2}}, \quad r = \|x\|, \quad N = 2. \tag{1.40}$$

In the three-dimensional case ($N = 3$), the fundamental solution is the singular GF—*a simple layer on the light cone* $\|x\| = ct$:

$$\hat{U}(x, t) = -\frac{1}{4\pi r}\delta(t - r/c) = \frac{1}{4\pi t}\delta(ct - r). \tag{1.41}$$

It defines a linear functional of the form:

$$\left(\hat{U}, \varphi\right) = -\left(\frac{\delta(t - r/c)}{4\pi r}, \varphi(x, t)\right) = -\frac{1}{4\pi}\int_{R^3} \frac{\varphi(x, r/c)}{r} dV(x). \tag{1.42}$$

On the light cone $\|x\| = ct$, fundamental solutions have a feature, in contrast to the features of fundamental solutions of Laplace equation, which are concentrated at zero, which is generally characteristic of elliptic equations.

Note that in the literature, as well as in V. Vladimirov [4, 5], the fundamental solution is written as

$$\hat{U}(x, t) = \frac{c}{4\pi}\delta\left(c^2t^2 - \|x\|^2\right),$$

which is not convenient for constructing solutions of the wave equation and boundary value problems for it. Representation (1.41) is preferable, since it often allows you to use the property of the delta function by a fixed x.

For $N > 1$ on moving wave fronts, the fundamental solutions are infinite, which fundamentally distinguishes the fundamental solutions of hyperbolic equations from elliptic ones. This does not allow directly to use the classical apparatus of potential theory to construct boundary integral equations for solving boundary value problems for hyperbolic equations.

To construct the integral representations of the solutions of wave equations in the three-dimensional case, we will further use the integral representations of complete and incomplete convolutions with simple layers on S and the light cone [12, 13], which can be proved using the definition of convolution in the space of generalized functions:

$$\beta(x,t)\delta_S(x)H(t)*f(x,t)H(t) = H(t)\int_0^t d\tau \int_S \beta(y,t-\tau)f(x-y,\tau)dS(y), \tag{1.43}$$

$$\alpha(x)\delta\left(t-\frac{\|x\|}{c}\right)*f(x,t)H(t) = H(t)\int_{r<ct} \alpha(y)f\left(x-y,t-\frac{\|y-x\|}{c}\right)dS(y), \tag{1.44}$$

$$\begin{aligned} &\alpha(x)\delta\left(t-\frac{\|x\|}{c}\right)*\beta(x,t)\delta_S(x)H(t) = \\ &= H(t)\int_{S_t(x)} \alpha(x-y)\beta\left(y,t-\frac{\|x-y\|}{c}\right)dS(y), \end{aligned} \tag{1.45}$$

$$\alpha(x)\delta\left(t-\frac{\|x\|}{c}\right)_x * \gamma(x)\delta_S(x) = \frac{\partial}{\partial t}\int_{S_t(x)} \alpha(x-y)\gamma(y)dS(y). \tag{1.46}$$

Here a variable under the convolution sign means incomplete convolution only for this variable, $S_t(x) = \{y \in S : \|x-y\| < ct\}$. Potentials (1.44, 1.45) are called *retarded potentials.*

1.7 Fundamental Solutions of Klein–Gordon–Fock Equation

The Klein–Gordon–Fock equations are an equation of quantum mechanics. It has the form:

$$\Box_c u + q(x)u = G(x,t), \quad x \in R^N, \quad t \in R^1. \tag{1.47}$$

In particular, the wave function $u(x,t)$ of a free relativistic particle of mass m_0 satisfies the homogeneous Klein–Gordon–Fock equation [14–16].

Hereinafter, we consider boundary value problems for a KGF equation of the form (1.47) under the assumption $q(x) = \pm m^2 = q_{\pm}$. To construct their solutions, we use fundamental solutions (1.47). Their form depends on the dimension of the space N [5, 6]: At $N = 1$.

for $q(x) = m^2$

$$\hat{U}(x,t) = -\frac{c}{2}H(ct - |x|)J_0\left(\frac{m}{2}\sqrt{c^2t^2 - |x|^2}\right), \tag{1.48}$$

for $q(x) = -m^2$:

$$\hat{U}(x,t) = -\frac{c}{2}H(ct - |x|)I_0\left(\frac{m}{2}\sqrt{c^2t^2 - |x|^2}\right), \tag{1.49}$$

where $J_0(...)$ is Bessel function of the first kind, $I_0(...)$ is Bessel function of an imaginary argument. At the front with $|x| = ct$ the fundamental decision has a gap: $[U]_{x=\pm ct} = \frac{c}{2}$.

Figure 1.1 shows the oscillogram of oscillations in the case (1.48) at fixed observation points (x^j) at $c = 1, m = 1$, $|x^j|$=1,10,50 on the interval $t \in [0; 100]$ (notations are introduced here: Uj$(t) = U(x^j, t)$. The graph clearly shows the jump and delay of the signal as the observation point moves away from the source.

At $N = 2$ $\hat{U}(x,t)$ is the regular GF:

for $q(x) = +m^2$:

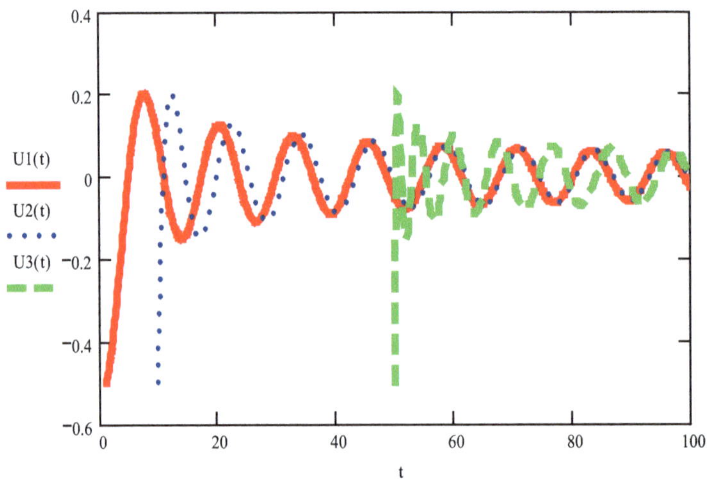

Fig. 1.1 Oscillogram of U at $t \in [0; 100]$

$$\hat{U} = \frac{H(ct-r)}{2\pi}\frac{ch\left(m\sqrt{c^2t^2-r^2}\right)}{\sqrt{c^2t^2-r^2}}, \quad r = \|x\| \tag{1.50}$$

for $q(x) = -m^2$:

$$\hat{U} = \frac{H(ct-r)}{2\pi}\frac{\cos\left(m\sqrt{c^2t^2-r^2}\right)}{\sqrt{c^2t^2-r^2}}. \tag{1.51}$$

The carrier $U(x, t)$ is a circle expanding in R^2 with a radius of $R = ct$. On its circumference—the wave front—$U(x, t)$turns to infinity:

$$U \sim \frac{1}{2\pi\sqrt{c^2t^2-r^2}} \quad \text{at} r \to ct - 0. \tag{1.52}$$

Figures 1.2 and 1.3 show the form of these solutions at a fixed time t.

For $N = 3$ $\hat{U}$ is singular GF. It has the following form.

for $q(x) = m^2$:

$$4\pi\hat{U}(x,t) = \frac{H(t)}{r}\delta(ct-r) - \frac{mH(ct-r)}{\sqrt{c^2t^2-r^2}}J_1\left(\frac{m}{c}\sqrt{c^2t^2-r^2}\right), \tag{1.53}$$

for $q(x) = -m^2$:

Fig. 1.2 $U(x,t)$ at $q(x) = m^2$

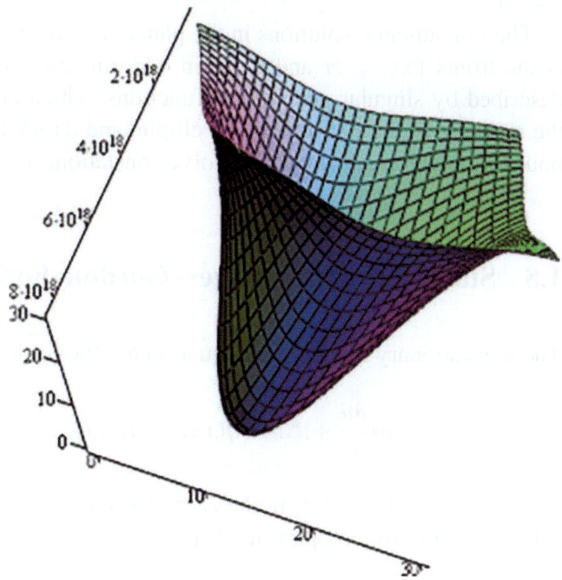

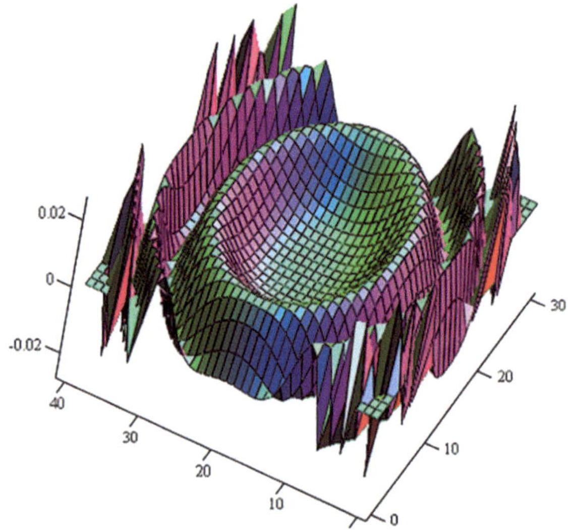

Fig. 1.3 $U(x,t)$ at $q(x) = -m^2$

$$4\pi \hat{U}(x,t) = \frac{H(t)}{r}\delta(ct-r) - \frac{mH(ct-r)}{\sqrt{c^2t^2-r^2}} I_1\left(\frac{m}{c}\sqrt{c^2t^2-r^2}\right). \tag{1.54}$$

The first term in (1.53, 1.54) is a simple layer on the light cone—a singular generalized function focused on the front $\|x\| = ct$. The second terms are regular GFs with a finite jump at the front $\|x\| = ct$, since $J(z) \sim 0.5z$, $I(z) \sim 0.5z$ at $z \to 0$.

The fundamental solutions in the planar and three-dimensional cases are infinite at the fronts $\|x\| = ct$ and those in odd-dimensional spaces ($N = 2m + 1$) are described by singular generalized functions, which complicates the application of the methods of the BIE theory for elliptic and parabolic equations for constructing boundary integral equations that solve nonstationary boundary value problems.

1.8 Stationary Schrodinger–Gordon–Fock Equation

The nonstationary Schrodinger equation has the form [16]:

$$i\alpha\frac{\partial u}{\partial t} + \Delta u + q(x)u = G(x,t), \quad x \in R^N, \ t \in R^1. \tag{1.55}$$

If $G(x,t) = f(x)e^{-i\omega t}$, in the case of stationary oscillations, it is reduced to the following form for complex amplitudes:

$$\Delta u^* + k^2 u^* + q(x)u^* = f(x), \quad x \in R^N, \tag{1.56}$$

where $k = \alpha\omega$.

We note that in this case the KGF equation has a similar form, where k is the wave number $k = \omega/c$. We call this *equation the stationary Schrodinger–Gordon–Fock equation.*

In the case when $q(x) = \pm m^2$, Eq. (1.56) for complex amplitudes has the form:

$$\Delta u^* + \left(k^2 \pm m^2\right)u^* = f(x). \tag{1.57}$$

For the upper sign, the equation coincides with the Helmholtz equation

$$\Delta u^* + \kappa^2 u^* = f^*(x). \tag{1.58}$$

Here $\kappa = \sqrt{k^2 + m^2} > 0$. For the bottom sign, setting $\kappa = \sqrt{\left|k^2 - m^2\right|}$ we have (1.58), only if $m < k$. At $m > k$ the equation transforms to

$$\Delta u^* - \kappa^2 u^* = f^*(x). \tag{1.59}$$

So, for $q(x) = -m^2$ we use fundamental solutions.

for $N = 2$:

$$\hat{U} = -\frac{i}{4}H_0^1(\kappa r) \ \text{if} m < k, \hat{U} = -\frac{i}{4}K_0(\kappa r) \ \text{if} m > k; \tag{1.60}$$

for $N = 3$:

$$\hat{U} = -\frac{e^{i\kappa r}}{4\pi r} \ \text{if} m < k, \hat{U} = -\frac{e^{-\kappa r}}{4\pi r} \ \text{if} m > k. \tag{1.61}$$

Here $K_0(\kappa r)$is the MacDonald function [17]. When $m > k$ the fundamental solution decays exponentially.

The property of fundamental solutions (1.22) allows to construct general solutions to all these equations for any generalized functions on the right-hand side that allow these convolutions. We will show these later on specific boundary value problems.

References

1. Schwartz, L.: Mathematical methods for physical sciences—Moskow "Mir" (1965)
2. Hörmander, L.: Linear differential operators with partial derivatives—Moskow "Mir" (1965)
3. Hermander, L.: Analysis of linear partial differential operators. In 4 volumes—Moskow "Mir" (1986)
4. Vladimirov, V.S.: Generalized functions in mathematical physics—Moskow "Nauka" (1979)
5. Vladimirov, V.S.: Equations of mathematical physics—Moskow "Nauka" (1981)
6. Vladimirov, V.S., Mikhailov, V.P., Vasharin, A.A., Karimova, Kh.H., Sidorov, Yu.V., Shabunin, M.I.: Collection of problems on the equations of mathematical physics—Moskow "Nauka" (1989)

7. Polyanin, A.D.: Handbook of linear equations of mathematical physics—Moskow“FIZ-MATLIT” (2001)
8. Sommerfeld, A.: Die Greensche Funktion der Schwingungsglerchung. Jber. Deutsch Math. Verein, Bd 21, pp. 309–353 (1912)
9. Atkinson, F.V.: On sommerfeld’s radiation condition. Philos. Mag. **40**, 645–651 (1949)
10. Kleinman, R.E., Wendland, W.: On Neumann’s method for the exterior Neumann problem for the Helmholtz equation. J. Math. Anal. Appl. **57**, 170–202 (1977)
11. Colton, D., Kleinman, R.E.: The direct and inverse scattering problems for an arbitrary cylinder. Dirichlet boundary conditions. In: Proceedings of Royal Society Edinburgh, v. 86A, pp. 2942 (1980)
12. Alexeyeva, L.A.: Generalized solutions of nonstationary boundary value problems for Maxwell’s equations. J. Comput. Math. Math. Phys. **42**(1), 76–88 (2002)
13. Alexeyeva, L.A.: Method of generalized functions in nonstationary boundary value problems for the wave equation. Math. J. **6**(1)(19), 1632
14. Kyurkchan, A.G.: On one scattering problem for the Helmholtz equation Reports of the USSR. Acad. Sci. **275**(1), 48–51 (1984)
15. Bogolyubov, N.N., Logunov, A.A., Todorov, I.T.: Fundamentals of the axiomatic approach in quantum field theory—Moskow “Nauka” (1970)
16. Bogolyubov, N.N., Shirkov, D.V.: Introduction to the theory of quantized fields—Moskow “Nauka” (1973)
17. Special Functions Reference. Abramowitz, M., StiganMoskow, I., (eds.) “Nauka” (1979)

Chapter 2
Integral Equations of Boundary Value Problems for Schrodinger–Gordon–Fock Equation: Scattering Amplitude

Among the most complex and relevant from the point of view of applications is the problem of wave diffraction by various inhomogeneities in continuous media (cavity, inclusion, crack, etc.), or when additional terms are introduced into the equation that include the scattering potential (the latter is typical for quantum mechanics problems). The determination of the diffraction pattern that occurs when waves are scattered by inhomogeneities leads to the solution of stationary and nonstationary boundary value problems of mathematical physics for equations of hyperbolic type depending on the type of incident wave. The theory of stationary wave diffraction is most developed when the frequency (length) of the incident wave, usually plane harmonic, is fixed. In quantum mechanics, such problems are called the scattering problem [1, 2]. In this case, the hyperbolic equations for the desired functions are transformed into elliptic for the complex amplitudes of the reflected waves, for the determination of which various methods are used. In the case when the scattering surface is canonical (circle, sphere, cylinder, plane), methods of complete and incomplete separation of variables are used to solve such problems. This class of problems has a very extensive bibliography for various continuous media and fields. For a diffuser of non-canonical form, depending on the type of diffuser, the method of integral equations or MBIE are used. To solve integral equations, analytically, as a rule, asymptotic methods are used that allow one to determine the far field from the scattering amplitude. The exact solution of such equations for determining the near field requires the solution of boundary integral equations.

This chapter discusses external stationary boundary value problems for the stationary elliptic Schrodinger–Gordon–Fock equation in a region bounded by a closed surface S on which an incident harmonic wave is diffracted. In addition, the equation contains a scattering potential $q(x)$, which also scatters the field of the incident wave.

In the space of generalized functions, analogues of the Gauss and Green formulas are constructed for the solutions of SGF equations of different dimensions, and their regular integral representations are given. Resolving integral equations are

L. Alexeyeva and A. Bayegizova, *Generalized Functions Method in Boundary Value Problems for Wave Equations*, Industrial and Applied Mathematics,
https://doi.org/10.1007/978-981-95-2802-8_2

constructed for three plane and three-dimensional boundary value problems. The scattering amplitude for the solutions of the third boundary value problem under the impedance boundary condition is determined. Examples of scattering amplitude for different scattering potentials and scattering surface are given.

The case of $q(x) = \pm m^2$ constants is considered, for which generalized solutions of boundary value problems of SGF equations and resolving singular boundary integral equations are constructed.

2.1 The Scattering Problem for the Stationary SGF Equation

We consider in the space of functions $C^2(R^3)$ the solutions of the stationary Schrödinger-Gordon-Fock (SGF) equation:

$$\Delta\varphi + k^2\varphi - q(x)\varphi = 0, \tag{2.1}$$

which has the form:

$$\varphi(\mathbf{k}, x) = \varphi^0 + \varphi^S = e^{i(\mathbf{k},x)} + \varphi^S(\mathbf{k}, x), \tag{2.2}$$

where φ^0 is the solution of Helmholtz equation:

$$\Delta\varphi^0(\mathbf{k}, x) + k^2\varphi^0(\mathbf{k}, x) = 0. \tag{2.3}$$

Here $\varphi^0(\mathbf{k}, x) = e^{i(\mathbf{k},x)}$ describe a plane harmonic wave moving in the direction of the wave vector $\mathbf{k}$, $\mathbf{k} = (k_1, k_2, k_3)$, $k = \|\mathbf{k}\|$, $x = (x_1, x_2, x_3)$, and scattering potential $q(x)$ is an integrable function with limited carrier (compact). The function φ^S describes scattered waves and must satisfy the Sommerfeld radiation conditions (1.31, 1.32) (depending on the dimension of space) at infinity.

When stating the direct scattering problem, it is assumed that the scattering potential $q(x)$ is known. It is required to find the scattering amplitude.

Theorem 1.1 *The solution to the scattering problem is asymptotically equivalent*

$$\varphi \sim e^{i(\mathbf{k},x)} - A(\mathbf{k}, e_x)\frac{e^{ik\|x\|}}{4\pi\|x\|} at\ \|x\| \to \infty,$$

where $A(\mathbf{k}, e_x)$—scattering amplitude—has the form:

$$A(\mathbf{k}, e_x) = q^*(\mathbf{k} - ke_x),$$

$q^*(\xi)$ is the Fourier transform $q(x)$, and $e_x = \frac{x}{\|x\|}$ is a unit vector in the direction of the observation point x.

Proof Substituting expression (2.2) into Eq. (2.1), we obtain

$$\Delta e^{i(\mathbf{k},x)} + k^2 e^{i(\mathbf{k},x)} - q(x)e^{i(\mathbf{k},x)} + (\Delta + k^2 - q(x))\varphi^S = 0,$$

where $(\mathbf{k}, x) = k_1x_1 + k_2x_2 + k_3x_3$. Since

$$\Delta e^{i(\mathbf{k},x)} + k^2 e^{i(\mathbf{k},x)} = 0,$$

then

$$(\Delta + k^2 - q(x))\varphi^S(k, x) = q(x)e^{i(k,x)}.$$

From here follows the expression

$$(\Delta + k^2)\varphi^S = q(x)e^{i(\mathbf{k},x)} + q(x)\varphi^S. \tag{2.4}$$

Let $U = \hat{U}_1$ be a fundamental solution to the Helmholtz Eq. (1.28). Then, using the property of the fundamental solution, we obtain

$$\varphi^S(\mathbf{k}, x) = U * q(x)e^{i(\mathbf{k},x)} + q(x)\varphi^S(\mathbf{k}, x) * U(x). \tag{2.5}$$

The integral form of this convolution depends on the dimension of the problem. Here we consider the three-dimensional case (N = 3), then

$$\hat{U} = \hat{U}_1 = -\frac{1}{4\pi\|x\|}e^{i(\mathbf{k},x)}.$$

Passing to the integral notation in (2.5), we obtain

$$\varphi^S = -\frac{1}{4\pi}\int\limits_{R^3} \frac{e^{ikr}}{r(x,y)} q(y)e^{i(k,y)}dV(y) - \frac{1}{4\pi}\int\limits_{R^3} \frac{q(y)e^{ikr}}{r(x,y)}\varphi^S(k,y)dV(y), \tag{2.6}$$

$$r(x, y) = \|x - y\| = r.$$

Equation (2.6) is a Fredholm equation of the second kind with a weakly polar core. The first term at a given potential $q(x)$ can be calculated for any x. Let's consider it in more detail:

$$-\frac{1}{4\pi}\int\limits_{R^3}\frac{e^{ik\|x-y\|}}{\|x-y\|}q(y)e^{i(\mathbf{k},y)}dV(y)=\Sigma_1$$

$$\|x-y\|=\sqrt{\|x\|^2-2\|x\|\|y\|\cos\theta+\|y\|^2}=\|x\|\sqrt{1-2\cos\theta\frac{\|y\|}{\|x\|}+\frac{\|y\|^2}{\|x\|^2}}.$$

Since supp $q(y)$ is a limited set, $\exists R_q>0$: $q(y)=0$ at $\|y\|>R_q$. Then

$$\Sigma_1(x)=-\frac{1}{4\pi\|x\|}\int\limits_{\text{supp }q}\frac{q(y)e^{i((\mathbf{k},y)+k\|x-y\|)}}{\sqrt{1-\frac{2\|y\|\cos\theta}{\|x\|}+\frac{\|y\|^2}{\|x\|^2}}}dV(y).$$

Similarly, the second term can be represented as

$$\Sigma_2(x)=-\frac{1}{4\pi}\int\limits_{\text{supp }q}\frac{q(y)\varphi^S(\mathbf{k},y)}{\|x-y\|}e^{ik\|x-y\|}dV(y)=$$

$$=-\frac{1}{4\pi\|x\|}\int\limits_{\text{supp }q}\frac{q(y)\varphi^S(\mathbf{k},y)}{\sqrt{1-2\cos\theta\frac{\|y\|}{\|x\|}+\frac{\|y\|^2}{\|x\|^2}}}e^{ik\|x-y\|}dV(y).$$

Fredholm integral equations of the second kind are solved by the method of successive approximations:

$$\varphi=K\varphi+\psi=K(K\varphi)+K\psi+\psi,$$

where K is integral operator. As a result, we obtain an asymptotic expansion in which we restrict ourselves to the asymptotic representation:

$$\Phi^S\sim\Phi_1^S=\psi,$$

$$\psi=\Sigma_1(x)=-\frac{1}{4\pi\|x\|}\int\limits_{\text{supp }q}\frac{q(y)e^{i((\mathbf{k},y)+k\|x-y\|)}}{\sqrt{1-\frac{2\|y\|\cos\theta}{\|x\|}+\frac{\|y\|^2}{\|x\|^2}}}dV(y),$$

$$K\psi=-\frac{1}{4\pi}\int\limits_{\text{supp }q}\frac{e^{ik\|x-y\|)}q(y)\sum_1(y)}{\|x-y\|}dV(y).$$

Using the Taylor series expansion formula, we obtain

$$\sqrt{1-2\cos\theta\frac{\|y\|}{\|x\|}+\frac{\|y\|^2}{\|x\|^2}}=\left(1-2\cos\theta\frac{\|y\|}{\|x\|}+\frac{\|y\|^2}{\|x\|^2}\right)^{1/2}=$$
$$=1-\frac{\|y\|}{\|x\|}\cos\theta+\frac{1}{2}\cdot\frac{\|y\|^2}{\|x\|^2}$$
$$+\frac{0.5(0.5-1)}{2!}\cdot\left(-2\frac{\|y\|}{\|x\|}\cos\theta+\frac{\|y\|^2}{\|x\|^2}\right)^2+\ldots=$$
$$=1-\cos\theta\, frac\|y\|\,\|x\|+O\left(\frac{R_q^2}{\|x\|^2}\right).$$

Therefore

$$\psi\sim-\frac{1}{4\pi\,\|x\|}\int\limits_{\text{supp}\,q}q(y)e^{i((\mathbf{k},y)+k\|x-y\|)}dV(y)+O\left(\frac{\|y\|}{\|x\|}\right).$$

We transform the integrand factor at $\|x\|\to\infty$ and $\|y\|\le R_q$:

$$e^{i((\mathbf{k},y)+k\|x-y\|)}=e^{i((k,y)}\exp\left(ik\,\|x\|\left\|e_x-\frac{y}{\|x\|}\right\|\right)$$
$$=e^{i((\mathbf{k},y)}e^{ik\|x\|}e^{-ik\|y\|\cos\theta}\exp\big(ikO\big(\|x\|^{-2}\big)\big).$$

Since $ikO\big(\|x\|^{-2}\big)\to 0$ at $\|x\|\to\infty$, then $\exp\big(ikO\big(\|x\|^{-2}\big)\big)\to 1$ and

$$e^{i((\mathbf{k},y)+k\|x-y\|)}\cong e^{i(\mathbf{k},y)}e^{ik\|x\|}e^{-ik\|y\|\cos\theta},$$

substituting, we obtain

$$\varphi_1^S\cong-\frac{e^{ik\|x\|}}{4\pi\,\|x\|}\int\limits_{\text{supp}\,q}q(y)e^{i(\mathbf{k},y)}e^{-ik\|y\|\cos\theta}dV(y). \tag{2.7}$$

We denote

$$A(\overrightarrow{k},e_x)=\int\limits_{\text{supp}\,q}q(y)e^{ik((k^0,y)-\|y\|\cos\theta)}dV(y),$$

where $\cos\theta=(e_x,y/\|y\|)$, $k^0=\frac{\mathbf{k}}{k}$ is unit wave direction vector, and $e_x=\frac{x}{\|x\|}$ is a unit vector in direction of an observation point x.

Then

$$\varphi_1^S=-\frac{e^{ik\|x\|}}{4\pi\,\|x\|}\int\limits_{\text{supp}\,q}q(y)e^{ik((k^0,y)-2(e_x,y))}dV(y)=-\frac{A(\mathbf{k},e_x)}{4\pi\,\|x\|}e^{ik\|x\|}.$$

The statement is proven.

Let us consider some examples of calculating the scattering amplitude.

Examples.

1. $q(x) = \prod_{j=1}^{3} H(a_j - |x_j|)$(*Parallelepiped*). Determine scattering amplitude $A(\overrightarrow{k}, e_x)$.

Solution. The Fourier transform of the scattering potential:

$$q^*(k) = \int_{\text{supp } q} q(y)e^{i(k,y)}dV(y) = \int_{-a_1}^{a_1}\int_{-a_2}^{a_2}\int_{-a_3}^{a_3} e^{i(k_1y_1+k_2y_2+k_3y_3)}dy_1dy_2dy_3 =$$

$$= \prod_{j=1}^{3}\frac{1}{ik_j}(e^{ik_ja_j} - e^{-ik_ja_j}) = 8\prod_{j=1}^{3}\frac{\sin(k_ja_j)}{k_j}.$$

The scattering amplitude, according to Theorem 1.1, is

$$q^*(\mathbf{k} - \|k\|e_x) = 8\prod_{j=1}^{3}\frac{\sin((k_j - \|k\|e_x)a_j)}{k_j - \|k\|e_x}.$$

We denote $b_j = k_j - \|k\|e_{x_j}$. At $e_x = k_0$ we have

$$q^*(k - \|k\|e_x) = g^*(0) = 8\prod_{j=1}^{3}\lim_{b_j\to 0}\frac{a_j\sin(b_ja_j)}{a_jb_j} = 8\prod_{j=1}^{3}a_j = 8a_1a_2a_3 = V,$$

i.e., equal to the volume of the carrier q.

2. $q(x) = H(a - \|x\|)$ (*Sphere*). Determine $A(\overrightarrow{k}, e_x)$.

Solution. We use spherical coordinates:$x_1 = \rho\sin\theta\cos\varphi$, $x_2 = \rho\sin\theta\sin\varphi$, $x_3 = \rho\cos\theta$, $0 \le \rho \le +\infty$, $0 \le \varphi \le 2\pi$, $0 \le \theta \le \pi$:

$$q^*(\mathbf{k}) = \int_{\|x\|<a} e^{i(\mathbf{k},x)}dV(x) = \int_0^{2\pi} d\varphi\int_0^{\pi}\sin\theta d\theta\int_0^{a} e^{ikR\cos\theta}R^2dR =$$

$$= \frac{8\pi}{3\|k\|^2}\left(\frac{\sin(ka)}{k} - a\cos(ka)\right)$$

$$A(\mathbf{k}, e_x) = \frac{8\pi}{3\|\mathbf{k} - ke_x\|^2}\left(\frac{\sin(a\|\mathbf{k} - ke_x\|)}{\|\mathbf{k} - ke_x\|} - a\cos(a\|\mathbf{k} - ke_x\|)\right).$$

3. A simple layer on a sphere of radius $R = a$: $g(x) = \delta(a - \|x\|)$. Find $A(\mathbf{k}, e_x)$.

Solution. Using spherical coordinates, we get

$$q^*(\mathbf{k}) = \int\limits_{\|x\|=a} e^{i(\mathbf{k},x)} dS(x) = -a^2 \int_0^{2\pi} d\varphi \int_0^{\pi} e^{ika\cos\theta} d(cos\theta) =$$

$$= -2\pi a^2 \left\{ \frac{e^{-ika} - e^{ika}}{ika} \right\} = 4\pi a \frac{\sin ka}{k} \quad \Rightarrow$$

$$A(\mathbf{k}, e_x) = 4\pi a \frac{\sin(a\|\mathbf{k} - ke_x\|)}{\|\mathbf{k} - ke_x\|}.$$

At $\mathbf{k} = ke_x$ here and in the second example the feature is disposable.

The formula of Theorem 1.1 determines the far field in the scattering of a harmonic wave.

2.2 Formulation of Stationary Boundary Value Problems for SGF Equation

We consider the external boundary value problem for the SGF equation. Let $S^- \subset R^N$ be the outer region with with respect to the closed region S^+ bounded by the closed Lyapunov surface S. We consider the solutions of the SGF equation in S^-:

$$\Delta u + k^2 u + q(x)u = f(x), \quad x \in S^-. \tag{2.8}$$

Let us consider three boundary value problems for this equation: at $x \in S$

$$\textit{Dirichlet problem } (\text{BVP I})\ u(x) = u_S(x), \tag{2.9}$$

$$\textit{Neumann problem } (\text{BVP II})\ \frac{\partial u}{\partial n} = p(x), \tag{2.10}$$

$$\textit{Roben problem } (\text{BVP III})\ ik\,u - \tfrac{\partial u}{\partial n} = g(x). \tag{2.11}$$

Here n is the outward unit normal to S. In the third boundary value problem, the expression on the left side is called *the impedance* [3, 4].

It is assumed that $u_S(x)$ belongs to the class of Gelder functions on S: $\exists\beta,\ 0 < \beta \leq 1$,

$$|u_S(x) - u_S(y)| < const\|x - y\|^{\beta} \tag{2.12}$$

for any $x \in S, y \in S, p(x) \in L_1(S),\ g(x) \in L_1(S)$.

For uniqueness of solutions of external boundary value problems, we use the Sommerfeld radiation conditions (1.31), (1.32), which uniquely determine the solution of external boundary value problems for Helmholtz Eq. (3).

Let us consider the *generalized* solutions of BVP on $D'(R^N)$. To construct it, we also use the fundamental solution of the Helmholtz equation $\hat{U}_1(x)$ (1.30).

Further for regular $\hat{U}$ we use the notation: $U(x, y) = \hat{U}(x - y)$. It is easy to see that the following symmetry properties are satisfied:

$$U(x, y) = U(y, x), \quad \frac{\partial U(x, y)}{\partial x} = -\frac{\partial U(y, x)}{\partial y}. \tag{2.13}$$

Theorem 2.1. *In $D'(R^N)$ the generalized solution $\hat{u}$ is represented as the convolution.*

$$\hat{u} = U(x) * f(x) H_S^-(x) - U(x) * q(x)\hat{u}(x) -$$
$$-U(x) * g(x)\delta_S(x) - \sum_{j=1}^{N} \frac{\partial U(x)}{\partial x_j} * u_s(x) n_j(x) \delta_S(x).$$

Proof. By Lemma 2.3, we have

$$\Delta\hat{u} + k^2\hat{u} + q(x)\hat{u} = f(x) - \frac{\partial u}{\partial n}\delta_S(x) - \sum_{j=1}^{N} \frac{\partial}{\partial x_j}\left\{u_S(x) n_j(x) \delta_S(x)\right\}. \tag{2.14}$$

As we see, the boundary values of the function and its normal derivative are the densities of the double and simple layers, respectively. According to the properties of the fundamental solution and the property of differentiation of convolution, solution (2.14) can be represented as convolution:

$$\begin{aligned} \hat{u}(x) = \hat{U}(x) * f(x) H_S^-(x) - \hat{U}(x) * q(x)\hat{u}(x) - \\ -\hat{U}(x) * p(x)\delta_S(x) - \sum_{j=1}^{N} \frac{\partial \hat{U}(x)}{\partial x_j} * u_S(x) n_j(x) \delta_S(x). \end{aligned} \tag{2.15}$$

Using linearity and the property of convolution differentiation, transforming the last term, we obtain the formula of the theorem.

The constructed representation of solutions is unique in the convolution algebra with $\hat{U}(x)$. Indeed, let there be two solutions $\hat{u}_1$ and $\hat{u}_2$. Then $\hat{u}_1 - \hat{u}_2 = q(x)(\hat{u}_1 - \hat{u}_2) * \hat{U}(x)$ is a solution to a homogeneous problem. Due to its uniqueness for external boundary value problems, $q(x)(\hat{u}_1 - \hat{u}_2) * \hat{U}(x) = 0$. But then, by virtue of the convolution property, this implies the uniqueness of a generalized solution: $\hat{u}_1 = \hat{u}_2$.

2.3 Integral Representations of Solutions of Boundary Value Problems

Theorem 2.1. *The solution of boundary value problems (2.8–2.10) satisfies an integral equation of the form.*

$$u(x)H_S^-(x) = \text{V.P.}\int_S \frac{\partial U(y,x)}{\partial n(y)} u_S(y)dS(y) - \int_S U(y,x)p(y)dS(y) +$$
$$+\int_{S^-} (f(y) - q(y)u(y))U(y,x)dV(y).$$

For $x \in S$ the first integral is singular, taken in the Value Principle sense.

Proof Due to the regularity of fundamental solutions, the formula is an integral representation of formula (2.15), where the last convolution, by virtue of the symmetry properties of U, is transformed

$$\sum_{j=1}^{N}\left(\frac{\partial U}{\partial x_j} * \left\{u(x)n_j(x)\delta_S(x)\right\}\right) =$$
$$= -\sum_{j=1}^{N}\int_S \frac{\partial U(y,x)}{\partial y_j} u(y)n_j(y)dS(y) = -\int_S \frac{\partial U(y,x)}{\partial n(y)} u(y)dS(y).$$

For $x \notin S$ integrands are continuous; all integrals exist and satisfy the Helmholtz equation. For $x \in S$ boundary points, the integral $\int_S \frac{\partial U(y,x)}{\partial n(y)} u(y)dS(y)$ does not exist since $\frac{\partial U}{\partial y_j}$, by virtue of (1.33) and (1.34), has the strong non-integrable order r^{-1} singularities for $N = 2$ and r^{-2} for $N = 3$. We show that, taking into account definition of $H_S^-(x)$ (1.13), the formula of the theorem is preserved if we take the integral in the sense of the principal value.

Let $x^* \in S$, $x \in S^-$. We regularize the divergent integral and let $x \to x^*$. Then

$$u(x^*) = \lim_{x\to x^*} u(x) = \lim_{x\to x^*}\int_{S^-} (f(y) - q(y)u(y))U(y,x)dV(y) -$$
$$-\lim_{x\to x^*}\int_S U(y,x)p(y)dS(y) + \lim_{x\to x^*}\int_S \left(u_S(y) - u(x^*)\right)\frac{\partial U(y,x)}{\partial n(y)}dS(y) +$$
$$+u(x^*)\lim_{x\to x^*}\int_S \frac{\partial U(y,x)}{\partial n(y)}dS(y).$$

All integrals, except the last, are also convergent, therefore

$$u(x^*) = \int_{S^-} (f(y) - q(y)u(y))U(y, x^*)dV(y) -$$

$$-\int_S U(y, x^*)p(y)dS(y) + \int_S (u_S(y) - u(x^*))\frac{\partial U(y, x^*)}{\partial n(y)}dS(y) +$$

$$+u(x^*)\left\{1 - k^2 \int_{S^-} U(y, x^*)dV(y)\right\}.$$

Here, to calculate the last integral, we used an analog of the Gauss formula (see Lemma 4.1). Since, by virtue of the same lemma,

$$\int_S \left(u_S(y) - u(x^*)\right)\frac{\partial U(y, x)}{\partial n(y)}dS(y) =$$

$$= \text{V.P.} \int_S u_S(y)\frac{\partial U(y, x)}{\partial n(y)}dS(y) - u(x^*)\left(\frac{1}{2} - k^2 \int_{S^-} U(y, x^*)dV(y)\right)$$

we have

$$u(x^*) = \int_{S^-} (f(y) - q(x)u(x))U(y, x^*)dV(y) -$$

$$-\int_S U(y, x^*)p(y)dS(y)) + \text{V.P.} \int_S u_S(y)\frac{\partial U(y, x^*)}{\partial n(y)}dS(y) -$$

$$-u(x^*)\left(\frac{1}{2} - k^2 \int_{S^-} U(y, x^*)dV(y)\right) + u(x^*)\left\{1 - k^2 \int_{S^-} U(y, x^*)dV(y)\right\}.$$

Moving $0.5u(x^*)$ to the left, taking into account the notation of the boundary functions (2.9, 2.10), we obtain the formula of the theorem for the boundary points.

The solution satisfies the conditions of Sommerfeld radiation at infinity due to the choice of the fundamental solutions of U and their asymptotic properties (1.31), (1.32), the boundedness of the contour S, and the carrier $q(x)$. The theorem is proven.

This theorem implies an integral representation of solutions for two- and three-dimensional spaces characteristic of mathematical physics problems, which are easy to write using the form of the fundamental solution $\hat{U}_1(x)$ (1.29), (1.30). For boundary value problems in the three-dimensional case, we obtain the representation that the following theorem gives.

Theorem 2.2. *The solution of three-dimensional boundary value problems satisfies the integral equation of the form:*

$$4\pi\,\widehat{u} = \int\limits_{S^-} (f(y) - q(y)u(y))\tfrac{e^{ikr}}{r(x,y)}dV(y) + \int\limits_{S} \left(iku(y) - \frac{\partial u}{\partial n(y)}\right)\tfrac{e^{ikr}}{r}dS(y) +$$

$$+\text{V.P.}\int\limits_{S} u(y)\tfrac{e^{ikr}}{r^2}\frac{\partial r}{\partial n(y)}dS(y), \quad r(x,y) = \|x-y\| = r,$$

where the last integral, singular for $x \in S$, is taken in the Value Principle sense.

This non-classical type equation resembles the Fredholm equation of the second kind in appearance, but it still contains surface integrals of the boundary function, the latter being singular. Here we find the asymptotic representation of the solutions of this equation for large x.

2.4 Scattering Amplitude of Stationary Boundary Value Problems

We consider the scattering problem taking into account the boundary conditions at $f(x) = 0, N = 3$. We seek the asymptotics of the solution of external boundary value problems at $\|x\| \to \infty$ in the form:

$$u(x) = e^{i(\mathbf{k},x)} - \frac{A(\mathbf{k}, e_x)}{r}e^{ikr} + o(r^{-1}), \quad e_x = \frac{x}{r}, \; r = \|x\|, \tag{2.16}$$

where the scattering amplitude $A(\mathbf{k}, e_x)$ is to be determined.

Theorem 2.3 *The scattering amplitude of the solutions of external BVP* (2.9–2.11) *for homogeneous SGF equation satisfying the Sommerfeld condition* (1.32) *has the form:*

$$4\pi A(\mathbf{k}, e_x) = q^*(\mathbf{k} - ke_x) + \int\limits_{S} \left(\frac{\partial u(y)}{\partial n(y)} - iku(y)\right)e^{ik(e_x,y)}dS(y),$$

where $q^*(\mathbf{k}) = \int\limits_{S^-} q(x)\exp(i(\mathbf{k},x))dx_1dx_2dx_3$.

Proof. Let θ be the angle between the vectors x, y, $r = \|x\|$. When $\|x\| \to \infty$ since [4]

$$\frac{1}{r} = \left(r^2 + \|y\|^2 - 2r\|y\|\cos\theta\right)^{-\frac{1}{2}} = r^{-1}\left\{1 + \sum_{m=1}^{\infty} P_m(\cos\theta)\left(\frac{\|y\|}{r}\right)^m\right\}$$

$$r^{-2} = 1/\left(r^2 + \|y\|^2 - 2r\|y\|\cos\theta\right) = \|x\|^{-2}\left(1 - \frac{\|y\|}{r}\cos\theta + O\left(\frac{\|y\|^2}{r^2}\right)\right),$$

where $P_m(z)$ are Legendre polynomials. Therefore

$$4\pi u(x) = -r^{-1}\int_{S^-} (q(y)u(y))e^{ikr}\left\{1 + \sum_{m=1}^{\infty} P_m(\cos\theta)\frac{\|y\|^m}{r^m}\right\}dV(y)+$$

$$+r^{-1}\int_{S} (g(y) - iku_S(y))e^{ikr}\left\{1 + \sum_{m=1}^{\infty} P_m(\cos\theta)\left\|\frac{y}{r}\right\|^m\right\}dS(y)+$$

$$+r^{-2}\int_{S} e^{ikr}\frac{\partial\|x-y\|}{\partial n(y)}u_S(y)\left(1 - \left\|\frac{y}{r}\right\|\cos\theta + O\left(\left\|\frac{y}{r}\right\|^2\right)\right)dS(y),$$

where we get

$$4\pi u(x) = r^{-1}\int_{S^-} (f(y) - q(y)u(y))e^{ik\|x-y\|}dV(y)+$$

$$+r^{-1}\int_{S} (p(y) - iku_S(y))e^{ik\|x-y\|}dS(y) + O\left(r^{-2}\right).$$

Since

$$e^{ik\|x-y\|} = \exp\left\{ikr\sqrt{1 - 2\left\|\frac{y}{r}\right\|\cos\theta + \left\|\frac{y}{r}\right\|^2}\right\} =$$

$$= \exp\left\{ikr\left(1 - \left\|\frac{y}{r}\right\|\cos\theta + O\left(\left\|\frac{y}{r}\right\|^2\right)\right)\right\} = e^{ikr}\exp\left(ik\|y\|\cos\theta + O(r^{-1})\right),$$

where $\exp\left(O(r^{-1})\right) = 1 + O(r^{-1})$. From these relations it follows that

$$u(x) = -\frac{e^{ikr}}{4\pi r}\int_{S^-} q(y)u(y)e^{ik\|y\|\cos\theta}dV(y)$$

$$+\frac{e^{ikr}}{4\pi r}\int_{S} (p(y) - iku_S(y))e^{ik\|y\|\cos\theta}dS(y) + o(r^{-1})$$

where we get the expression for the scattering amplitude:

$$4\pi A(\mathbf{k}, e_x) =$$
$$= -\int\limits_{S^-} q(y)u(y)e^{ik\|y\|\cos\theta} dV(y) + \int\limits_{S} (p(y) - iku_S(y))e^{ik\|y\|\cos\theta} dS(y).$$

Taking into account the introduced notation and Theorem 3.2, we obtain the formula of the theorem. This theorem implies.

Theorem 2.4. *The scattering amplitude of boundary value problem III is.*

$$4\pi A(e_x) = q^*(\mathbf{k} - ke_x) + F_{ke_x}\big[g(x)\delta_S(x)\big].$$

Proof. In view of Theorem 2.3 and boundary condition (2.11) we obtain the required.

The formula of this theorem allows us to determine the far field for solving the third boundary value problem without solving the corresponding integral equation. Here are some examples of determining the scattering amplitude for particular types of scattering potential and the shape of the scatterer.

Examples. S is a sphere of radius a, and the impedance on S is constant: $g(y) = const = I$. Find $A(\overrightarrow{k}, e_x)$ for different $q(x)$.

(a) $q(x) = \frac{1}{\|x\|} H(\|x\| - a)H(b - \|x\|)$.

We calculate by passing to spherical coordinates,

$$q^*(\mathbf{k}) = \int\limits_{a<r<b} r^{-1} e^{i(\mathbf{k},x)} dV(x) = -\int_0^{2\pi} d\varphi \int\limits_a^b r dr \int_0^{\pi} e^{ikr\cos\theta} d(cos\theta) =$$
$$= 4\pi \int\limits_a^b r \sin kr dr = -\frac{4\pi}{k} \int\limits_a^b r d(\cos kr) = -\frac{4\pi}{k}\left(r\cos kr|_a^b - \int\limits_a^b \cos kr dr \right) =$$
$$= -\frac{4\pi}{k}\left(b\cos kb - a\cos ka - \frac{\sin kb - \sin ka}{k} \right), \quad r = \|x\|.$$

Using the formula of Theorem 2.4, we obtain the scattering amplitude:

$$A(\mathbf{k}, e_x) = Ia\frac{\sin(ak)}{k} - \frac{4\pi}{\chi}\left(b\cos\chi b - a\cos\chi a - \frac{\sin\chi b - \sin\chi a}{\chi} \right),$$

where $\chi = \|\mathbf{k} - ke_x\|$.

(b) $q(x) = \frac{H(r-a)}{r}$,

$$q^*(\mathbf{k}) = \int\limits_{r>a} r^{-1} e^{i(\mathbf{k},x)} dV(x) =$$

$$= -\int_0^{2\pi} d\varphi \int_0^{\pi} e^{ika\cos\theta} d(cos\theta) \int\limits_a^{\infty} rdr = 2\pi \frac{\sin ka}{k} (r^2)\big|_a^{\infty} = \infty.$$

That is, the classical Fourier transform does not exist. However, note that

$$q(x) = \frac{H(r-a)}{r} = \frac{1}{r} - \frac{H(a-r)}{r}.$$

From the first example it is easy to get

$$F_\xi\left[\frac{H(a-r)}{r}\right] = \frac{4\pi}{\|\xi\|}\left(a\cos\|\xi\|a - \frac{\sin\|\xi\|a}{\|\xi\|}\right)$$

and since

$$\Delta\left(-\frac{1}{4\pi\|x\|}\right) = \delta(x) \Rightarrow \quad \|\xi\|^2 F_\xi\left[\frac{1}{\|x\|}\right] = 4\pi \Rightarrow F_\xi\left[\frac{1}{\|x\|}\right] = \frac{4\pi}{\|\xi\|^2}.$$

Then

$$q^*(\mathbf{k}) = F_{\mathbf{k}}\left[\frac{1}{\|x\|}\right] - F_{\mathbf{k}}\left[\frac{H(a-\|x\|)}{\|x\|}\right] = \frac{4\pi}{k}\left(1 + a\cos ka - \frac{\sin ka}{k}\right)$$

$$A(\mathbf{k}, e_x) = Ia\frac{\sin(ak)}{k} + \frac{4\pi}{\chi}\left(1 + a\cos\chi a - \frac{\sin\chi a}{\chi}\right), \text{ where} \chi = \|\mathbf{k} - ke_x\|.$$

(c) $q(x) = H(r-a)e^{-\lambda r}, \quad \lambda > 0.$

$$q^*(\mathbf{k}) = \int\limits_{\|x\|>a} e^{i(\mathbf{k},x)-\lambda\|x\|} dV(x) = \int\limits_0^{2\pi} d\varphi \int\limits_0^{\pi} d\theta \int\limits_a^{\infty} e^{ikr\cos\theta - \lambda r} r^2 dr =$$

$$= \frac{4\pi}{k}\int\limits_a^{\infty} e^{-\lambda r}\sin(kr) rdr = \frac{4\pi}{k^3}\int\limits_{ak}^{\infty} re^{-\lambda r/k}\sin(r)dr,$$

$$q^*(\mathbf{k} - ke_x) = \frac{4\pi}{\|\mathbf{k} - ke_x\|^3} \int\limits_{a\|\mathbf{k}-ke_x\|}^{\infty} re^{-\lambda r/\|\mathbf{k}-ke_x\|}\sin(r)dr,$$

$$A(\mathbf{k}, e_x) = aI\frac{\sin(ak)}{k} + \frac{1}{\|\mathbf{k} - ke_x\|^3} \int\limits_{a\|\mathbf{k}-ke_x\|}^{\infty} re^{-\lambda r/\|\mathbf{k}-ke_x\|}\sin(r)dr.$$

2.5 Analogue of Green Formula and Singular Boundary Integral Equations of Boundary Value Problems for the SGF Equation

In the case when the scattering potential is constant, the solution of boundary value problems can be reduced to boundary integral equations.

Theorem 2.5. *In $D'(R^N)$ $(N = 2, 3)$, a generalized solution of the SGF equation for $q(x) = -m^2$ is represented as*

$$u(x)H_S^-(x) = \text{V.P.}\int_S \frac{\partial U(y,x)}{\partial n(y)}u(y)dS(y) - \int_S U(y,x)\frac{\partial u}{\partial n(y)}dS(y) + \\ + \int_{S^-} f(y)U(y,x)dV(y), \tag{2.17}$$

where $U(x)$ is determined by formulas (1.60), (1.61) *depending on the dimension of the value m.*

Proof By Lemma 2.3, we have

$$\Delta\hat{u} + \big(k^2 + q(x)\big)\hat{u} = f(x) - \frac{\partial u}{\partial n}\delta_S(x) - \sum_{j=1}^{N}\frac{\partial}{\partial x_j}\big\{u_S(x)n_j(x)\delta_S(x)\big\}. \tag{2.18}$$

According to the properties of the fundamental solution and the property of differentiation of convolution, solution (2.18) can be represented as a convolution:

$$\hat{u} = \hat{U}(x) * f(x)H_S^-(x) - \hat{U}(x) * p(x)\delta_S(x) - \sum_{j=1}^{N}\frac{\partial \hat{U}(x)}{\partial x_j} * u_S(x)n_j(x)\delta_S(x) \tag{2.19}$$

where the fundamental solution $U(x)$ is a regular generalized function and it is determined for $N = 2$ by formulas (1.60) depending on the value of the difference $(k^2 - m^2)$, for $N = 3$—by formulas (1.61). Using linearity and the property of convolution differentiation, transforming the last term, we obtain the formula of the theorem.

Let's prove the formula on the boundary of the region. Let $x \in S^-$, $x^* \in S$, and $x \to x^*$.

$$\lim_{x\to x^*} u(x) = u(x^*) =$$

$$= \lim_{x\to x*} \int_S \frac{\partial U(y,x)}{\partial n(y)}(u(y)-u(x^*))dS(y) + u(x^*) \lim_{x\to x^*} \int_S \frac{\partial U(y,x)}{\partial n(y)} dS(y)$$

$$- \lim_{x\to x^*} \int_S U(y,x)\frac{\partial u}{\partial n(y)} dS(y) + \lim_{x\to x^*} \int_{S^-} f(y)U(y,x)dV(y) =$$

$$= \int_S \frac{\partial U(y,x^*)}{\partial n(y)}(u(y)-u(x^*))dS(y) + u(x^*)\left(1-k^2\int_{L^-} U(x,y)dV(y)\right) -$$

$$- \int_S U(y,x^*)\frac{\partial u}{\partial n(y)} dS(y) + \int_{S^-} f(y)U(y,x^*)dV(y) =$$

$$= \text{V.P.} \int_S u(y)\frac{\partial U(y,x^*)}{\partial n(y)} dS(y) - u(x^*)\text{V.P.} \int_S \frac{\partial U(y,x^*)}{\partial n(y)} dS(y) +$$

$$+u(x^*)\left(1-k^2\int_{L^-} U(x,y)dV(y)\right) - \int_S U(y,x^*)\frac{\partial u}{\partial n(y)} dS(y) + \int_{S^-} f(y)U(y,x^*)dV(y) =$$

$$= \text{V.P.} \int_S u(y)\frac{\partial U(y,x^*)}{\partial n(y)} dS(y) - u(x^*)\left(0,5-k^2\int_{L^-} U(x,y)dV(y)\right) +$$

$$+u(x^*)\left(1-k^2\int_{S^-} U(x^*,y)dV(y)\right) - \int_S U(y,x^*)\frac{\partial u}{\partial n(y)} dS(y) + \int_{S^-} f(y)U(y,x^*)dV(y) =$$

$$= \text{V.P.} \int_S u(y)\frac{\partial U(y,x^*)}{\partial n(y)} dS(y) + 0,5u(x^*) - \int_S U(y,x^*)\frac{\partial u}{\partial n(y)} dS(y) +$$

$$+ \int_{S^-} f(y)U(y,x^*)dV(y).$$

Here, when going over to the limit, we use the Gelder property of the boundary values of the function u and the equality of the eigen integrals to their principal values. Moving $0,5u(x^*)$ to the left side, we obtain the formula of the theorem for boundary points.

The constructed representation of solutions is unique. Indeed, let there be two solutions of (2.18) satisfying the same boundary conditions u_1 and u_2. Then

$$\Delta(\hat{u}_1 - \hat{u}_2) + \left(k^2 + q(x)\right)(\hat{u}_1 - \hat{u}_2) = 0$$

is a solution to a homogeneous problem. Due to its uniqueness for external boundary value problems for the Helmholtz equation: $\hat{u}_1 - \hat{u}_2 = 0$. But then $u_1 = u_2$.

We formulate a particular form of this theorem for plane and three-dimensional problems in the form of the following theorems.

Theorem 2.6. *The solution of the boundary value problems of the SGF-equation for $q(x) = -m^2$ ($N = 2$) is representable in the form:*

at $m < k$

$$4i\,u(x)H_S^-(x) = \int\limits_{S^-} f(y)H_0(\kappa r)dV(y) - \int\limits_S H_0(\kappa r)\frac{\partial u}{\partial n(y)}dS(y) - \\ -\kappa \text{V.P.}\int\limits_S H_1(\kappa r)u(y)\frac{\partial r}{\partial n(y)}dS(y), \tag{2.20a}$$

at $m > k$

$$4iu(x)H_S^-(x) = \\ = \int\limits_{S^-} f(y)K_0(\kappa r)dV(y) - \int\limits_S K_0(\kappa r)\frac{\partial u}{\partial n(y)}dS(y) \\ -\kappa \text{V.P.}\int\limits_S K_1(\kappa r)u(y)\frac{\partial r}{\partial n(y)}dS(y), \tag{2.20b}$$

where $\kappa = \sqrt{\left|k^2 - m^2\right|}$.

Proof We write Theorem 5.1 using the form of fundamental solutions. Since

$$\frac{\partial}{\partial n(y)}H_0(\kappa r) = \kappa H_0{\prime}(\kappa r)\frac{\partial r}{\partial n(y)} = -\kappa H_1(\kappa r)\frac{\partial r}{\partial n(y)},$$

$$\text{V.P.}\int\limits_S \frac{\partial U(y,x)}{\partial n(y)}u(y)dS(y) = \text{V.P.}\int\limits_S \frac{i\kappa}{4}H_1(\kappa r)\frac{\partial r}{\partial n(y)}u(y)dS(y).$$

Similarly, we calculate

$$\frac{\partial}{\partial n(y)}K_0(\kappa r) = \kappa K_0{\prime}(\kappa r)\frac{\partial r}{\partial n(y)} = -\kappa K_1(\kappa r)\frac{\partial r}{\partial n(y)}.$$

Substituting, we obtain the second formula of the theorem.

Theorem 2.7. *The solution of the SGF-equation at* $q(x) = -m^2,\ N = 3$, *is representable*

at $m < k$

$$4\pi u(x)H_S^-(x) = \int\limits_S \frac{e^{i\kappa r}}{r}\left(\frac{\partial u}{\partial n(y)} + i\kappa u(y)\frac{\partial r}{\partial n(y)}\right)dS(y) - \\ -\text{V.P.}\int\limits_S \frac{e^{i\kappa r}}{r^2}\frac{\partial r}{\partial n(y)}u(y)dS(y) - \int\limits_{S^-} f(y)\frac{e^{i\kappa r}}{r}dV(y) \tag{2.21a}$$

at $m > k$

$$4\pi u(x)H_S^-(x) = -\int\limits_S \frac{e^{-\kappa r}}{r}\left(\frac{\partial u}{\partial n(y)} + \kappa u(y)\frac{\partial r}{\partial n(y)}\right)dS(y) - \\ -\text{V.P.}\int\limits_S \frac{e^{-\kappa r}}{r^2}\frac{\partial r}{\partial n(y)}u(y)dS(y) - \int\limits_{S^-} f(y)\frac{e^{-\kappa r}}{r}dV(y), \tag{2.21b}$$

where $\kappa = \sqrt{|k^2 - m^2|}$.

Proof. We write Theorem 5.1 using the form of the fundamental solutions (1.61_1) and (1.61_2). Because

$$\frac{\partial}{\partial n(y)}\left(\frac{e^{i\kappa r}}{r}\right) = \frac{e^{i\kappa r}}{r}\left(i\kappa - \frac{1}{r}\right)\frac{\partial r}{\partial n(y)},$$

$$\text{V.P.}\int\limits_S \frac{\partial U(y,x)}{\partial n(y)}u(y)dS(y) = -\text{V.P.}\int\limits_S \frac{e^{i\kappa r}}{r}\left(i\kappa - \frac{1}{r}\right)\frac{\partial r}{\partial n(y)}u(y)dS(y) =$$
$$= -i\kappa\int\limits_S \frac{e^{i\kappa r}}{4\pi r}\frac{\partial r}{\partial n(y)}u(y)dS(y) + \text{V.P.}\int\limits_S \frac{e^{i\kappa r}}{4\pi r^2}\frac{\partial r}{\partial n(y)}u(y)dS(y).$$

Substituting, we obtain the first formula of the theorem. Similarly, we get

$$\frac{\partial}{\partial n(y)}\left(\frac{e^{-\kappa r}}{r}\right) = -\frac{e^{-\kappa r}}{r}\left(\kappa + \frac{1}{r}\right)\frac{\partial r}{\partial n(y)}$$

and further (2.21_2). The theorem is proven.

In the case when $q(x) = m^2$, the theorems are formulated similarly to Theorems 2.6 and 2.7, only the first formulas of the theorems are used, where $\kappa = \sqrt{k^2 + m^2}$.

The formula of the theorems allows you to find unknown boundary functions, solving the corresponding BIE, and then restore the solution in the domain of definition. When $x \in S$ these formulas give singular BIE for solving the Neumann problem.

Due to the unique solvability of the external Neumann and Dirichlet boundary value problems for the Helmholtz equation (see Theorems 3.13, 3.21, and 3.25, p. 89.95.97 [3]), a solution to these boundary equations exists and is unique.

The solution of the Neumann problem for the GHF equation can be obtained using these formulas by the method of successive approximations due to the compactness of the boundary integral operators on the class of Gelder functions [3].

References

1. Kyurkchan, A.G.: On one scattering problem for the Helmholtz equation//Reports of the USSR. Acad. Sci. **275**(1), 48–51 (1984)
2. Lax, P., Phillips, R.: Scattering theory—Moskow "Mir" (1971)
3. Colton, D., Kress, R.: Methods of integral equations in scattering theory—Moskow "Nauka", p. 311 (1987)
4. Vladimirov, V.S.: Equations of mathematical physics—Moskow "Nauka", p. 512 (1981)

Chapter 3
Nonstationary Boundary Value Problems for D'Alembert Wave Equation and Their Solutions

Here, the method of generalized functions for constructing solutions of initial-boundary value problems for the wave equation, the d'Alembert equation in spaces of different dimensions, is described. Dynamic analogues of Green's formulas are constructed for solving boundary value problems for the wave equation in the space of generalized functions. Their regular integral representations and singular boundary integral equations are constructed to solve the corresponding initial-boundary value problems.

3.1 Generalized Solutions of D'Alembert Equation: Shock Waves

Let us consider a multidimensional analogue of the d'Alembert equation

$$\Box_c u \equiv \Delta u - c^{-2}\frac{\partial^2 u}{\partial t^2} = G(x,t), \quad x \in R^N, t \in R^1. \tag{3.1}$$

Here $G(x, t)$ is a locally integrable function, $\Box_c$ is the wave operator (d'Alembertian), and Δ is the Laplace operator (Laplacian).

Equation (3.1) is strictly hyperbolic; the class of its solutions contains functions discontinuous in derivatives [1]. Fracture surfaces F in R^{N+1} are characteristic surfaces of Eq. (3.1) that satisfy the characteristic equation in space $R^{N+1}=\{(x, \tau \equiv ct)\}$:

$$\nu_\tau^2 - \sum_{j=1}^{N} \nu_j^2 = 0, \tag{3.2}$$

L. Alexeyeva and A. Bayegizova, *Generalized Functions Method in Boundary Value Problems for Wave Equations*, Industrial and Applied Mathematics,
https://doi.org/10.1007/978-981-95-2802-8_3

where $\nu(x, \tau) = (\nu_1, \ldots, \nu_{N,} \nu_\tau)$ is a normal vector to F. It corresponds to a cone of characteristic normals which is a *light* cone for which $\nu_\tau = \nu_{N+1} < 0$. In R^N such surfaces move with unit velocity along τ:

$$1 = -\nu_\tau / \|\nu\|_N, \quad \|\nu\|_N = \sqrt{\nu_j \nu_j} \tag{3.3}$$

(summation from 1 to N is carried out everywhere over the repeated indices i, j in the derivative hereinafter). In a space R^N they correspond to *wave fronts* (F_t) moving with speed c in time t. On them the Hadamard continuity conditions are satisfied [2]:

$$[u(x,t)]_{F_t} = 0, \qquad [\dot{u} + cn_i\, u,_i]_{F_t} = 0, \tag{3.4}$$

where with $[f(x,t)]_{F_t}$ the jump f to F_t is designated:

$$[f(x,t)]_{F_t} = f^+(x,t) - f^-(x,t) = \lim_{\varepsilon \to +0} (f(x + \varepsilon n, t) - f(x - \varepsilon n, t)), x \in F_t.$$

$n(x,t)$ is the normal unit vector to F_t, directed toward the propagation of the wave front:

$$n_i = \frac{\nu_i}{\|\nu\|_N} = \frac{\text{grad}\, F_t}{\|\text{grad}\, F_t\|}, \qquad i = 1, \ldots, N. \tag{3.5}$$

The last equality is true if the equation of the wave front can be represented in the form $F_t(x,t) = 0$ under the condition of existence of *grad* F_t.

The class of such solutions of hyperbolic equations is called *shock waves*; on their fronts the derivatives of functions and even the functions themselves can suffer jumps.

From the second condition (3.4) follows conditions on the fronts:

$$\dot{u}^- + cn_i\, u,_i^- = \dot{u}^+ + cn_i\, u,_i^+ . \tag{3.4*}$$

If in front of the wave front $u \equiv 0$ (medium at rest), this equality gives a useful relation at the wave front:

$$(\text{grad}\, u, n) = -c^{-1}\dot{u}, \quad x \in F_t.$$

Note that the tangent derivatives to the characteristic surface, due to the continuity of u, are also continuous:

$$[u,_\tau \gamma_\tau + u_j \gamma_j]_F = 0 \text{ for } \forall \gamma : (\nu, \gamma) = 0. \tag{3.6}$$

In particular, if $\gamma = \gamma^j = (-\nu_j, \nu_\tau \delta_1^j, \nu_\tau \delta_2^j, \nu_\tau \delta_2^j)$, this leads to conditions of the form:

$$\left[-u_{,\tau}\,\nu_j + u_j\nu_\tau\right]_F = 0 \quad\Rightarrow\quad \left[\dot{u}n_j + cu_j\right]_{F_t} = 0. \tag{3.6*}$$

Next, we consider functions $u(x,t)$ that are continuous together with derivatives up to the second order inclusive almost everywhere, with the exception of a finite or countable number of discontinuity surfaces—wave fronts that are sufficiently smooth almost everywhere on which the conditions for jumps (3.3) are satisfied. We call such solutions *classical*. We show that they are *generalized solutions* of Eqs. (3.1).

For this, we consider Eq. (3.1) on the space of generalized functions $D'(R^{N+1})$.

Definition A function $\hat{f} \in D'(R^N)$ is called *a generalized solution* of Eq. (3.1) if for any $\varphi \in D(R^{N+1})$ equality holds.

$$(\Box_c\hat{f},\varphi) \equiv (\hat{f},\Box_c\varphi) = (G,\varphi), \quad \forall\varphi \in D\left(R^{N+1}\right).$$

Lemma 3.1 *If* $u(x,t)$ *is the classical solution* (3.1), *then* $\hat{u}(x,t)$ *is its generalized solution.*

Proof. If u has a finite discontinuity on F, then in $D'(R^{N+1})$ accordance with the rules for determining the generalized derivative,

$$\hat{u}_{,j} = u_{,j} + [u]_F\nu_j\delta_F(x,\tau),$$

where the first term on the right is the classical derivative with respect to x_j, $(j = 1, \ldots, N+1)$, $\|\mathbf{v}\| = 1$, and δ_F is a simple layer on F. It is a singular generalized function:

$$\left([u]_F\nu_j\delta_F(x,\tau),\varphi(x,\tau)\right) = \int\limits_F [u(x,\tau)]_F\nu_j(x,\tau)\varphi(x,\tau)dF(x,t)$$

for $\forall\varphi \in D(R^{N+1})$. Here, the integral is surface with respect to F. Due to the continuity of u outside the wave front,

$$[u]_F = \lim_{\varepsilon\to+0}(u(x+\varepsilon n,t) - u(x-\varepsilon n,t)) = u^+(x,t) - u^-(x,t) = [u]_{F_t}.$$

Therefore, taking into account (3.5), we obtain

$$\hat{u}_{,j} = u_{,j} + [u]_{F_t}\nu_j\delta_F(x,\tau) = u_{,j} + \|\nu\|_N[u]_{F_t}n_j\delta_F(x,\tau), \quad j = 1,\ldots,N;$$
$$\hat{u}_{,jj} = u_{,jj} + \left[u_{,j}\right]_{F_t}\|\nu\|_N n_j\delta_F(x,\tau) + \partial_j\left\{\|\nu\|_N[u]_{F_t}n_j\delta_F(x,\tau)\right\}, \quad j = 1,\ldots,N.$$

By virtue of (3.3),

$$\hat{u}_{,\tau} = u_{,\tau} + [u]_{F_t}\nu_\tau\delta_F(x,\tau) = c^{-1}u_{,t} - \|\nu\|_N[u]_{F_t}\delta_F(x,\tau),$$
$$\hat{u}_{,tt} = c^{-2}u_{,tt} - c^{-1}[u_{,t}]_{F_t}\|\nu\|_N\delta_F(x,\tau) - \partial_\tau\left\{\|\nu\|_N[u]_{F_t}\delta_F(x,\tau)\right\}.$$

Given these equalities and the Hadamard conditions (3.4), we get

$$\Box_c \hat{u} = \Box_c u + \left\{c^{-1}[u_{,t}]_{F_t} + \left[n_j u_{,j}\right]_{F_t}\right\} \|v\|_N \delta_F(x,\tau) +$$
$$c^{-1}\partial_t\left\{\|v\|_N [u]_{F_t} \delta_F(x,\tau)\right\} + \partial_j\left\{\|v\|_N [u]_{F_t} n_j \delta_F(x,\tau)\right\} = \hat{G}(x,t), \tag{3.7}$$
$$j = 1, ..., N.$$

Since all the densities of simple and double layers on F_t are equal to zero. Indeed, the second term is equal to zero due to the second condition on the fronts. In addition, the other two terms, by virtue of (3.4), are equal 0, since its effect on φ is defined as

$$\left(c^{-1}\partial_t\left\{\|v\|_N [u]_{F_t} \delta_F(x,\tau)\right\} + \partial_j\left\{\|v\|_N [u]_{F_t} n_j \delta_F(x,\tau)\right\}, \varphi(x,t)\right) =$$
$$= \left(\|v\|_N [u]_{F_t} \delta_F(x,\tau), \frac{\partial \varphi}{\partial n} - \varphi_{,\tau}\right).$$

Remark 1 From this lemma it follows that the conditions on the fronts of shock waves are easily obtained by considering the classical solutions of hyperbolic equations as generalized. It is enough to equate to zero the densities of the corresponding independent singular generalized functions—analogues of simple, double, and other layers that arise during the generalized differentiation of solutions. The determination of such conditions on the basis of classical methods is a very laborious procedure.

Remark 2 Equation (3.1) admits generalized solutions with a jump of derivatives on moving surfaces $F(x,t) = 0$, the speed of which can depend on the front point:

$$v(x,t) = -\frac{F_{,t}}{\|\text{grad}\, F\|},$$

if the Hadamard conditions (3.4) are satisfied on them. Such solutions can be generated by the right-hand side of the equation if the carrier $G(x,t)$ expands over time in R^N.

3.2 Statement of Nonstationary Boundary Value Problems: Uniqueness of Solutions

Let us suppose that in the region $x \in S^-$ bounded by the Lyapunov surface S ([3], p. 409), a solution $u(x,t)$ is constructed of the wave equation for $t \geq 0$. We denote $D^- = S^- \times R^+$, $S^- \in R^N$, $R^+ = [0,\infty)$, and $n = (n_1, ..., n_N)$ is a unit vector of the outer normal to S.

Initial conditions. At $t = 0$

$$u(x,0) = u_0(x) \text{ for } x \in S^- + S,\; u_{,t}(x,0) = \dot{u}_0(x) \text{ for } x \in S^-. \tag{3.8}$$

Here we consider two boundary value problems corresponding to the Dirichlet and Neumann boundary conditions:

$$\text{BVP I } u(x,t) = u_S(x,t) \text{ for } x \in S, \quad t \geq 0 \tag{3.9}$$

$$\text{BVP II } \frac{\partial u}{\partial n} = p(x,t) \text{ for } x \in S, \quad t \geq 0. \tag{3.10}$$

For the first boundary value problem, the conditions for matching the boundary and initial data are satisfied:

$$u_0(x) = u_S(x,0) \text{ for } x \in S. \tag{3.11}$$

At wave fronts, if they arise, the Hadamard conditions (3.4) are satisfied.

Note that shock waves always arise if the condition for matching the initial and boundary data on the velocities is not fulfilled (hereinafter we denote $u_{,t} = \dot{u}$):

$$\dot{u}_0(x) = \dot{u}_S(x,0) \text{ for } x \in S, \tag{3.12}$$

which is typical for physical tasks. In this case, at the initial moment of time, a shock wave front forms at the boundary *S*, which propagates at a speed of *c* in S^-. To construct continuously differentiable solutions, this condition is necessary. We won't enter it here.

It is assumed that the initial conditions are given and one of the boundary conditions corresponding to the considered boundary value problem is known.

We introduce *the energy functions*

$$E = 0,5\left(u_{,\tau}^2 + \sum_{j=1}^{N} u_{,j}^2\right), \quad L = 0,5\left(u_{,\tau}^2 - \sum_{j=1}^{N} u_{,j}^2\right),$$

which in physical problems corresponds to the total energy density and the Lagrange function of the scalar field *u*(*x*,*t*).

Lemma 3.2 *If* $u(x,t)$ *is the classical solution of* (3.1), *then at the fronts of the shock waves the following conditions for the jumps E and L are satisfied:*

$$[E]_{F_t} = -c^{-1}\left[\dot{u}\frac{\partial u}{\partial n}\right]_{F_t}, \quad [L(x,t)]_{F_t} = c^{-2}\left(\dot{u}^- + c\frac{\partial u^-}{\partial n}\right)[\dot{u}]_{F_t}. \tag{3.13}$$

Proof. By virtue of the easily proved equality:

$$[ab] = a^+[b] + b^-[a], \tag{3.14}$$

taking into account (3.4) and (3.6*), we obtain

$$\begin{aligned}
\left[cE+\dot{u}\frac{\partial u}{\partial n}\right]&=\left[0,5\left(c^{-1}\dot{u}^2+cu,_j u,_j\right)+\dot{u}\frac{\partial u}{\partial n}\right]=\\
&=0,5\left[\dot{u}\left(c^{-1}\dot{u}+\frac{\partial u}{\partial n}\right)\right]+0,5\left[cu,_j u,_j+\dot{u}\frac{\partial u}{\partial n}\right]=\\
&=0,5c^{-1}\left(\dot{u}^+\left[\dot{u}+c\frac{\partial u}{\partial n}\right]+\left(\dot{u}^-+cu,_j^- n_j\right)[\dot{u}]\right)+\\
&+0,5u,_j^+\left[cu,_j+\dot{u}n_j\right]+0,5\left(cu,_j^-+\dot{u}^-n_j\right)\left[u,_j\right]=\\
&=0,5c^{-1}[\dot{u}]\left(\dot{u}^-+cu,_j^- n_j\right)+0,5\left[u,_j\right]\left(cu,_j^-+\dot{u}^-n_j\right)=\\
&=0,5cu,_j^-\left[u,_j+c^{-1}n_j\dot{u}\right]+0,5c^{-1}\dot{u}^-\left[cn_ju,_j+\dot{u}\right]=0
\end{aligned}$$

(here n is a normal to the shock front in R^N). This implies the formula (3.13) of the lemma.

Further, since $[a^2]=\left(a^++a^-\right)[a]$, and by virtue of (3.4) and (3.6), we obtain

$$\begin{aligned}
[L]=0,5\left[u,_\tau^2-\sum_{j=1}^{N}u,_j^2\right]&=0,5\left(u,_\tau^++u,_\tau^-\right)\left[u,_\tau\right]-0,5\left(u,_j^++u,_j^-\right)\left[u,_j\right]=\\
&=0,5\left(u,_\tau^++u,_\tau^-\right)\left[u,_\tau\right]+0,5\left(u,_j^++u,_j^-\right)n_j\left[u,_\tau\right]=\\
&=0,5\left\{\left(u,_\tau^++n_ju,_j^+\right)+\left(u,_\tau^-+n_ju,_j^-\right)\right\}\left[u,_\tau\right]=c^{-2}\left(\dot{u}^-+c\frac{\partial u^-}{\partial n}\right)[\dot{u}].
\end{aligned}$$

This is the formula of Lemma (3.14). The lemma is proven.

Remark 1 Condition (3.13) can easily be obtained by considering the corresponding equation in $D'(R^{N+1})$:

$$\begin{aligned}
\hat{E},_\tau-(u,_\tau u,_j),_j&=-u,_\tau G+\left\{[E]\nu_\tau-\left[u,_\tau u,_j\right]\nu_j\right\}\delta_F(x,\tau)=\\
&=-u,_\tau G-\|\nu\|_N\left\{[E]+\left[u,_\tau u,_j\right]n_j\right\}\delta_{F_t}(x,t).
\end{aligned}$$

Therefore, for (3.13) to be satisfied in $D'(R^{N+1})$, it is necessary that

$$[E]_F+\left[u,_\tau u,_j\right]_F n_j=\left[E+c^{-1}\dot{u}\frac{\partial u}{\partial n}\right]_F=0.$$

Remark 2 Note that if in front of the wave front $u\equiv 0$, then, taking into account (3.4*), we have $[L(x,t)]_{F_t}=0$, i.e., in this case the function L is continuous.

Theorem 3.1 *If u(x,t) is a classical solution of a boundary value problem, then.*

$$\int\limits_{S^-} (E(x,t) - E(x,0))dV(x) =$$

$$= \int\limits_0^t \int\limits_S (\dot{u}_S(x,t)p(x,t))dS(x)dt - \int\limits_0^t dt \int\limits_{D^-} G(x,t)u_{,t}\, dV(x).$$

Proof. Multiplying (3.1) by $u_{,\tau}$ in the domain of differentiability, after simple transformations we get

$$E_{,\tau} - (u_{,\tau}\, u_{,j})_{,j} = -u_{,\tau}\, G. \tag{3.15}$$

Now we integrate (3.15) over the region D^-, taking into account the partition of the integration region by wave fronts F_k. We note that the first two terms can be considered as the divergence of the corresponding vector in R^{N+1}, which is continuous in the regions between the fronts. Therefore, using the Ostrogradsky–Gauss theorem in R^{N+1}, we obtain

$$\int\limits_{D^-} E_{,\tau} dV(x)d\tau - \int\limits_{D^-} (u_{,\tau}\, u_{,j})_{,j}\, dV(x)d\tau + \int\limits_{D^-} u_{,\tau}\, G(x,\tau)dV(x)d\tau =$$

$$= \int\limits_{D^-} u_{,\tau}\, G(x,\tau)dV(x)d\tau + \int\limits_{S^-} (E(x,\tau) - E(x,0))dV(x) - \int\limits_0^\tau \int\limits_S (u_{,\tau}\, u_{,j}\, n_j)dS(x)d\tau +$$

$$+ \sum_{F_k} \int\limits_{F_k} [E\nu_\tau - u_{,\tau}\, u_{,j}\, \nu_j]_{F_k} dF_k(x,\tau) = 0$$

where $dF_k(x,\tau)$ is the differential of the surface area at the corresponding point of the wave front. By virtue of (3.3) and (3.13),

$$\left[E\nu_\tau - u_{,\tau}\, u_{,j}\, \nu_j\right]_{F_k} = -\frac{\|\nu\|_N}{c}\left[cE + \dot{u}\frac{\partial u}{\partial n}\right] = 0.$$

Therefore, the last integral is zero. Given the notation for boundary functions, we obtain the formula of the theorem. Its corollary is the following theorem.

Theorem 3.2 *If a classical solution to the first (second) boundary value problem exists, then it is unique.*

Proof. Due to the linearity of the problem, it is sufficient to prove the uniqueness of the zero solution. For it, $G = 0$, and the initial and corresponding boundary conditions are zero. Then it is easy to see that from Theorem 2.1 it follows:

$$\int\limits_{S^-} E(x,t)dV(x) = 0.$$

Since E is non-negative, therefore, $E \equiv 0 \Rightarrow u = const$. From the initial conditions it follows $u \equiv 0$.

Theorem 3.3 *If $u(x,t)$ is the classical solution of the boundary value problem, then*

$$\int_{D^-} L(x,t)dV(x)dt = \int_{D^-} u(x,t)G(x,t)dV(x)dt-$$

$$-\int_0^t \int_S (u_S(x,t)p(x,t))dS(x)dt + c^{-2}\int_{S^-} (u\dot{u}(x,t) - u_0\dot{u}_0(x))dV(x).$$

Proof. Multiplying (3.1) by $u(x,t)$, after simple transformations we get

$$L(x,\tau) + (uu,_j),_j - (uu,_\tau),_\tau = G. \tag{3.16}$$

We integrate (3.13) over D^-, taking into account the partition by its wave fronts F_k. Similarly, as in Theorem 2.1, using the Ostrogradsky–Gauss theorem, we have

$$\int_{D^-} L\,dV(x)d\tau = \int_{D^-} uGdV(x)d\tau + \int_{D^-} \big((uu,_\tau),_\tau - (uu,_j),_j\big)dV(x)d\tau =$$

$$= \int_{D^-} uG\,dV(x)d\tau + c^{-1}\int_{S^-} (uu,_t - u_0\dot{u}_0)dV(x) - \int_0^\tau d\tau \int_S uu,_j n_j dS(x)+$$

$$+\sum_{F_k}\int_{F_k} \big[uu,_\tau \nu_\tau - uu,_j \nu_j\big]_{F_k} dF_k(x,\tau).$$

Due to the conditions of Hadamard,

$$\int_{F_k} \big[uu,_\tau \nu_\tau - uu,_j \nu_j\big]_{F_k} dF_k(x,\tau) = \int_{F_k} u\big[u,_\tau \nu_\tau - u,_j \nu_j\big]_{F_k} dF_k(x,\tau) = 0.$$

Therefore, taking into account the conditions on the boundary (3.9, 3.10), we obtain the formula of the theorem.

3.3 Dynamic Analogue of Green's Formula in the Space of Generalized Functions

To construct a solution to boundary value problems, we pass to the space $D'(R^{N+1})$. For this, we introduce the characteristic function of the solution domain $H_D^-(x,t) \equiv H_S^-(x)H(t)$, where $H_S^-(x)$ is the characteristic function of the set S^- equal to 0.5 at its boundary S, and $H(t)$ is the Heaviside function equal to 0.5 at $t = 0$. H_D^- is the characteristic function of the space–time cylinder $D^- = \{S^- \times R_+\}$. It is easy to show that its partial derivatives are equal to

$$\frac{\partial H_D^-}{\partial x_j} = -n_j(x)\delta_S(x)H(t), \quad \frac{\partial H_D^-}{\partial t} = -n_j(x)H_S^-(x)\delta(t). \tag{3.17}$$

We introduce regular generalized functions

$$\hat{u}(x,t) = u(x,t)H_D^-(x,t), \hat{G}(x,t) = G(x,t)H_S^-(x)H(t),$$

where $u(x,t)$ is the classical solution of a boundary value problem that determines the solution and the right-hand side by zero outside the domain of definition.

Let us consider the action of the wave operator on this function $\hat{u}(x,t)$. Since the gap

$$[u]_S = 0 - u(x,t)|_{x\in S} = -u_S(x,t),$$

performing generalized differentiation, similar to that in p. 1.1, and using (3.1) in the domain of differentiability, we obtain

$$\begin{gathered}\Box_c\hat{u} = -\frac{\partial u}{\partial n}\delta_S(x)H(t) - H(t)\big(un_j\delta_S(x)\big)_{,j} - \\ -c^{-2}H_S^-(x)u_0(x)\dot{\delta}(t) - c^{-2}H_S^-(x)\dot{u}_0(x)\delta(t) + \hat{G}(x,t),\end{gathered} \tag{3.18}$$

where $\beta(x,t)\delta_S(x)H(t)$ is a simple layer on the lateral surface of the cylinder $\{S \times R^+\}$, $\delta(t)$ is the Dirac function, and $\frac{\partial u}{\partial n}$ is the normal derivative of n by S. Note that the densities of simple and double layers here are determined by given initial conditions and boundary conditions, some of which, depending on the boundary value problem being solved, are known, and part is not.

The solution to Eq. (3.18) is the convolution of the fundamental solution to the equation with the right-hand side of Eq. (3.1). As such, we take a fundamental solution $\hat{U}(x,t)$ satisfying the equation:

$$\Box_c U = \delta(x)\delta(t), \tag{3.19}$$

and *radiation conditions*:

$$\hat{U}(x,t) = 0 \text{ at } t < 0; \quad \hat{U}(x,t) = 0 \text{ at } \|x\| > ct. \tag{3.20}$$

The first condition is the zero solution in the absence of disturbances up to the initial time, and the second condition is related to the finite velocity of propagation of disturbances from the origin, which is equal to c.

This function is called *the Green function* of Eq. (3.1). It describes the waves coming from the origin. If it is known, then solution (3.1) can be represented as the sum of functional convolutions:

$$\begin{aligned} \hat{u}(x,t) &= -\hat{U}(x,t) * \frac{\partial u}{\partial n}\delta_S(x)H(t) - \left(\hat{U} * un_j\delta_S(x)H(t)\right)_{,j} \\ &-c^{-2}\left(\hat{U} \underset{x}{*} H_S^-(x)u_0(x)\right)_{,t} - c^{-2}\hat{U} \underset{x}{*} H_S^-(x)\dot{u}_0(x) + \hat{G} * \hat{U}, \end{aligned} \tag{3.21}$$

where the symbol "$\underset{x}{*}$" means incomplete convolution only in x, since when taking the convolution we can use the property of the δ-function. Moreover, the solution (3.21) is unique in the class of functions that can be in convolution with U, and it satisfies radiation conditions of the type (3.20), the form of which depends on the region of definition $\hat{G}(x,t)$ and the initial conditions.

We write (3.21) for zero initial data and $G = 0$, using the convolution differentiation property:

$$\hat{u}(x,t) = -\hat{U}(x,t) * \frac{\partial u}{\partial n}\delta_S(x)H(t) - \hat{U}_{,j} * un_j\delta_S(x)H(t). \tag{3.22}$$

This formula expresses the solution of the boundary value problem in terms of the boundary values of the desired function and its normal derivative and is similar to the Green formula for the solutions of the Laplace equation. However, due to the peculiarities of the fundamental solutions of hyperbolic equations at the wave front, the form of which depends on the dimension of space, its integral representation gives diverging integrals (in the second term). To construct its regular integral representation, we introduce the following generalized functions—antiderivatives in t:

$$\hat{W}(x,t) = \hat{U}(x,t) * \delta(x)H(t) = \hat{U}(x,t) \underset{t}{*} H(t), \quad \hat{W}_{,t} = \hat{U}(x,t), \tag{3.23}$$

$$\hat{H}(x,m,t) = \frac{\partial \hat{W}(x,t)}{\partial x_i}m_i = \frac{\partial \hat{W}(x,t)}{\partial m}.$$

Note that due to the boundedness of the carrier with respect to t on the left, both convolutions exist. It is easy to see that they are also solutions of (3.1) for

$$G = H(t)\delta(x) \text{ and} G = H(t)\frac{\partial \delta(x,t)}{\partial m}$$

respectively.

Theorem 3.4 *In $D'(R^{N+1})$ the solution of the nonstationary boundary value problem satisfies the equation:*

$$
\begin{aligned}
\hat{u}(x,t) = -\hat{U}(x,t) * \frac{\partial u}{\partial n}\delta_S(x)H(t) - \left(\hat{W}_{,j} * \dot{u}n_j\delta_S(x)H(t)\right) - \\
-\hat{W}_{,j} \underset{x}{*} u_0(x)n_j(x)\delta_S(x) + c^{-2}\hat{U} \underset{x}{*} H_S^-(x)\dot{u}_0(x) = \\
+c^{-2}\left(\hat{U} \underset{x}{*} H_S^-(x)u_0(x)\right)_{,t} + \hat{G} * \hat{U}.
\end{aligned} \tag{3.24}
$$

At zero initial data and $G = 0$

$$
\hat{u}(x,t) = -\hat{U}(x,t) * \frac{\partial u}{\partial n}\delta_S(x)H(t) - \hat{W}_{,j} * \dot{u}n_j\delta_S(x)H(t). \tag{3.25}
$$

Proof**.** It is easy to show, using the definition of the derivative of a generalized function and the continuity of u, that

$$
\left(un_j\delta_S(x)H(t)\right)_{,t} = \dot{u}(x,t)n_j(x)\delta_S(x)H(t) + u(x,0)n_j(x)\delta_S(x)\delta(t).
$$

Using this equality and the property of convolution differentiation, we have

$$
\begin{aligned}
\left(\hat{U} * un_j\delta_S(x)H(t)\right)_{,j} = \left(\hat{W}_{,t} * un_j\delta_S(x)H(t)\right)_{,j} = \\
\hat{W}_{,j} * \left(un_j\delta_S(x)H(t)\right)_{,t} = \hat{W}_{,j} * \left(un_j\delta_S(x)H(t)\right)_{,t} = \\
= \hat{W}_{,j} * \dot{u}(x,t)n_j(x)\delta_S(x)H(t) + \hat{W}_{,j} * u(x,0)n_j(x)\delta_S(x)\delta(t).
\end{aligned}
$$

Since

$$
\hat{W}_{,j} * u(x,0)n_j(x)\delta_S(x)\delta(t) = \hat{W}_{,j} \underset{x}{*} u_0(x)n_j(x)\delta_S(x),
$$

substituting these relations in (3.22), we obtain the formula of the theorem.

It follows from the theorem that the solution to the problem is completely determined by the boundary values of the normal derivative of the function $u(x,t)$ and its speed $\dot{u} = u_{,t}$.

By analogy with the representation of solutions of the Laplace equation, these formulas can be called *a dynamic analogue of the Green formula.* Formula (3.25) has an advantage over (3.22), since it allows us to go directly to its integral notation without regularizing the integrand functions at the fronts.

For $x \in S$, this formula gives, as we show below, a boundary integral equation for solving the boundary value problem, if for $x \in S$ either

$$
\text{BVP I} \quad \dot{u}(x,t) = v(x,t), \tag{3.26}
$$

or

$$\text{BVP II} \quad \frac{\partial u}{\partial n} = p(x,t) \tag{3.27}$$

are known. Then solving the BIE at the boundary, we find the second, after which we can find $u(x,t)$ by formula (3.25).

3.4 Dynamic Analogue of Gauss Formula in the Space of Generalized Functions

We introduce the shifts functions

$$U(x,y,t) = \hat{U}(x-y,t), \quad W(x,y,t) = \hat{W}(x-y,t),$$

$$H(x,y,m,t) = \hat{H}(x-y,m,t),$$

which, by virtue of the symmetry properties of the wave operator and the δ-function, satisfy the following symmetry relations:

$$U(x,y,t) = U(y,x,t), \quad W(x,y,t) = W(y,x,t), \quad \frac{\partial W}{\partial x_j} = -\frac{\partial W}{\partial y_j}, \tag{3.28}$$

$$H(x,y,m,t) = -H(y,x,m,t) = -H(x,y,-m,t). \tag{3.29}$$

Lemma 3.3 (dynamic analogue of the Gauss formula). *In $D'(R^{N+1})$ the characteristic function of the domain of definition has the following representation*:

$$H_D^-(x,t) = -\hat{W}_{,i} \underset{x}{*} n_i(x)\delta(x) - c^{-2}\Big(\hat{U} \underset{x}{*} H_S^-(x)\Big)_{,t}. \tag{3.30}$$

Proof. We collapse both sides of Eq. (3.19) for $\hat{U}$ with $H_S^-(x)H(t)$ using the properties of convolution differentiation (3.20):

$$-U_{,j} * n_j(x)\delta_S(x)H(t) - c^{-2}U_{,t} * H_S^-(x)\delta(t) = H_D^-(x,t).$$

Performing a convolution with respect to *t*, taking into account (3.23), we obtain the formula of the lemma.

Formula (3.29) is an analogue of the well-known Gauss formula for the double-layer potential (p. 406, [3]), which gives an integral representation of the characteristic function of the set using the fundamental solution of the Laplace equation.

The Gauss formula is often used to derive the BIE of boundary value problems for elliptic equations and systems. Similarly, you can use the dynamic analogue of the Gauss formula to derive the BIE of hyperbolic equations. Here, to construct the BIE, we use a dynamic analogue of Green's formula.

The integral record of these relations depends on the dimension of the problem. Next, we give an integral representation of the formulas of Theorem 2.1 and Lemma 3.3 for spaces of dimension $N = 1,2,3$ most characteristic of physical problems.

3.5 The Solution of Plane Nonstationary Boundary Value Problems: Singular Boundary Integral Equations

At first let us consider the problem with zero initial conditions in the case N = 2. We use next notations: $dS(y)$ is the differential of the arc length of the curve S at the point y, $dV(y) = dy_1 dy_2$, $S_t(x) = \{y \in S, \quad r = \|y - x\| < ct\}$, $S_t^-(x) = \left\{y \in S^-, \ r < ct\right\}$, $r = \|x - y\|$.

Theorem 3.5 *For $N = 2$, a function $\hat{u}(x, t)$ satisfying the initial zero conditions $(u_0 = 0, \ \dot{u}_0 = 0)$ has the following integral representation:*

(first form)

$$\hat{u}(x,t) = \frac{c}{2\pi} \int_0^t d\tau \int_{S_\tau(x)} \left(\frac{\partial u(y, t-\tau)}{\partial n(y)} + \frac{1}{r} \frac{\partial r}{\partial n(y)} \tau \dot{u}(y, t-\tau) \right) \frac{ds(y)}{\sqrt{c^2\tau^2 - r^2}},$$

which, when changing the order of integration, has the form:
(second form)

$$\hat{u}(x,t) = \frac{c}{2\pi} \int_{S_t(x)} dS(y) \int_{r/c}^{t} \frac{\partial u(y, t-\tau)}{\partial n(y)} \frac{d\tau}{\sqrt{c^2\tau^2 - r^2}} + \\ + \frac{c}{2\pi} \int_{S_t(x)} \frac{1}{r} \frac{\partial r}{\partial n(y)} dS(y) \int_{r/c}^{t} \frac{\tau \dot{u}(y, t-\tau) d\tau}{\sqrt{c^2\tau^2 - r^2}}. \tag{3.31}$$

For $x \in S$, the second integral on the right is singular, taken in the sense of the principal value.

Proof. For $N = 2$ Green's function has the following form:

$$\hat{U}(x,t) = -\frac{cH(ct - \|x\|)}{2\pi\sqrt{c^2t^2 - \|x\|^2}}. \tag{3.32}$$

Computing by formulas (3.23), we find $\hat{W}$ and $\hat{H}$

$$\hat{W}(x,t) = -\frac{H(ct - \|x\|)}{2\pi} \ln\left(\frac{ct - \sqrt{c^2t^2 - \|x\|^2}}{R}\right),$$
$$\hat{H}(x,m,t) = -\frac{ct\,H(ct - \|x\|)}{2\pi\sqrt{c^2t^2 - \|x\|^2}} \frac{(x,m)}{\|x\|^2}. \tag{3.33}$$

If we write down the convolutions in (3.24) in the integral form, considering these representations, we obtain the relations of the theorem. Note that for $x \notin S$ the integrals on the right are regular, therefore for such x integrals on the right and left are regular functions.

Let us prove the validity of (3.31) for $x \in S$, taking into account the definition of $H_S^-(x)$. Denote the ε-neighborhood of the point x ($\varepsilon << ct, t > 0$):

$$Ш_\varepsilon(x) = \{y : r < \varepsilon\}, \quad S_\varepsilon^-(x) = S^- - Ш_\varepsilon(x),$$

$$S_\varepsilon^+(x) = S^+ - Ш_\varepsilon(x), \quad \Gamma_\varepsilon(x) = \{y \in S : r \le \varepsilon\}, \quad S_\varepsilon = S - \Gamma_\varepsilon,$$

$$Ш_\varepsilon^- = S^- \cap Ш_\varepsilon, \quad Ш_\varepsilon^+ = S^+ \cap Ш_\varepsilon,$$

$$\Gamma_\varepsilon^\pm(x) = \{y \in S^\pm : r = \varepsilon\}, \quad \frac{\partial r}{\partial n(y)} = r_{,j}\, n_j(y), \quad r = \|y - x\|, \quad r_{,j} = \frac{\partial r}{\partial y_j} = \frac{y_j - x_j}{r}.$$

We transform the contour S in a neighborhood of the point x, by passing it along the ε-semicircle in S^- ($\varepsilon < < ct$, $t > 0$). We write a dynamic analogue of Green's formula for the contour $S_\varepsilon + \Gamma_\varepsilon^-$:

$$0 = \frac{c}{\pi} \int\limits_{S_\varepsilon(x)+\Gamma_\varepsilon^-(x)} H(ct - r)dS(y) \int\limits_{r/c}^{t} \frac{\partial u(y, t-\tau)}{\partial n(y)} \frac{d\tau}{\sqrt{c^2\tau^2 - r^2}} +$$
$$+ \frac{c}{2\pi} \int\limits_{S_\varepsilon(x)} H(ct - r)\frac{1}{r}\frac{\partial r}{\partial n(y)} dS(y) \int\limits_{r/c}^{t} \frac{\tau \dot{u}(y, t-\tau)d\tau}{\sqrt{c^2\tau^2 - r^2}} +$$
$$+ \frac{c}{2\pi} \int\limits_{\Gamma_\varepsilon^-(x)} H(ct - r)\frac{1}{r}\frac{\partial r}{\partial n(y)} dS(y) \int\limits_{r/c}^{t} \frac{\tau \dot{u}(y, t-\tau)d\tau}{\sqrt{c^2\tau^2 - r^2}}.$$

At $\varepsilon \to 0$, the first integral, due to the weak singularity of the integrand, tends to the integral over S, the second integral tends to the integral in the sense of the principal value, which also exists, since the integrand has a singularity of the form $1/r$ and contains the function $\frac{\partial r}{\partial n(y)} = \frac{(n(y), y-x)}{r}$. The latter is asymptotically equivalent $\frac{\partial r}{\partial n(x)} = \frac{(n(x), y-x)}{r}$ for $y \to x$ which is antisymmetric with respect to the point x. The integral factor at $r \to 0$ does not have a singularity. Let us consider the last term, which we denote by $J_\varepsilon(x)$. On Γ_ε^- $r = \varepsilon$, $\frac{\partial r}{\partial n(y)} = -\frac{\partial r}{\partial r} = -1$, $dS(y) = \varepsilon\, d\theta$, where θ is a polar angle with vertex at x; θ_1 and θ_2 are the angles of the ends Γ_ε^-, numbered in order when going around the Γ_ε contour in the positive direction. Given these ratios

$$J_\varepsilon(x) = \frac{c}{2\pi}\int\limits_{\theta_1}^{\theta_2}\frac{\varepsilon}{\varepsilon}d\theta\int\limits_{\varepsilon/c}^{t}\frac{\tau\dot{u}(y, t-\tau)d\tau}{\sqrt{c^2\tau^2 - r^2}} = \frac{(\theta_2-\theta_1)}{2\pi}\int\limits_{\varepsilon/c}^{t}\frac{\tau\dot{u}(y, t-\tau)d\tau}{\sqrt{\tau^2-(r/c)^2}}.$$

Passing to the limit on $\varepsilon \to 0$, we obtain

$$\lim_{\varepsilon\to 0}(\theta_2-\theta_1) = -\pi,$$

$$\lim_{\varepsilon\to 0} J_\varepsilon(x) = -0,5\int\limits_0^t \dot{u}(y, t-\tau)d\tau = -0,5u(x, t).$$

Transferring this term to the left side, taking into account the definition $H_D^-(x, t) = 0,5$ for $x \in S$, we obtain the validity of the formula on the boundary. Since regular and generalized functions are left and right in (3.31), by virtue of the Du Bois-Reymond lemma, the equality valid in the class of generalized functions is also valid in the usual sense. The theorem is proven.

Relation (3.31) gives a singular boundary integral equation for solving boundary value problems.

Theorem 3.6 *For $N = 2$, the solution of the boundary value problem satisfies a singular boundary integral equation of the form:*

$$\pi u_S(x, t) = \int\limits_{S_t(x)} dS(y)\int\limits_{r/c}^{t}\frac{\partial u(y, t-\tau)}{\partial n(y)}\frac{d\tau}{\sqrt{\tau^2-(r/c)^2}} +$$
$$+V.P.\int\limits_{S_t(x)}\frac{1}{r}\frac{\partial r}{\partial n(y)}dS(y)\int\limits_{r/c}^{t}\frac{\tau\dot{u}_S(y, t-\tau)d\tau}{\sqrt{\tau^2-(r/c)^2}}, \quad x \in S.$$

In the case of the first boundary value problem, the left-hand side of the equation and the second integral on the right are known, determined by the boundary condition, and the first integral contains a kernel with a weak singularity, but not at a point, as is usual for elliptic problems, but at the front of Green's function. Solving it, we determine the normal derivative of the desired function on the boundary, after which formula (3.31) allows us to calculate the solution at any point in the domain of definition. In the case of the second boundary value problem, we have a singular BIE for determining the unknown boundary functions of the unknown function u. Solving it, we determine its values at the boundary, after which (3.31) we find the solution.

In the case of nonzero initial conditions, the solution of the problem is given by the following theorem.

Theorem 3.7 *For $N = 2$, the solution of the initial-boundary value problem has the following integral representation:*

$$2\pi \hat{u}(x,t) = \int_0^t d\tau \int_{S_\tau(x)} \left(\frac{\partial u(y,t-\tau)}{\partial n(y)} + \frac{\tau}{r}\frac{\partial r}{\partial n(y)}\dot{u}(y,t-\tau) \right) \frac{dS(y)}{\sqrt{\tau^2-(r/c)^2}} +$$
$$+c^{-1}\frac{\partial}{\partial t}\int_{S_t^-(x)} \frac{u_0(y)dV(y)}{\sqrt{c^2t^2-r^2}} - c^{-1}\int_{S_t^-(x)} \frac{\dot{u}_0(y)dV(y)}{\sqrt{c^2t^2-r^2}} +$$
$$+\int_0^t cd\tau \int_{S_\tau(x)} \frac{G(y,t-\tau)dV(y)}{\sqrt{c^2\tau^2-r^2}} - \int_{S_t(x)} \frac{ctu_0(y)}{r}\frac{\partial r}{\partial n(y)}\frac{dS(y)}{\sqrt{c^2\tau^2-r^2}}.$$

Proof follows from Theorem 5.1 and of the form (3.32), if we write convolutions with the initial data in integral form. Here the integrals from the second to the fourth coincide with the Poisson formula for the Cauchy problem for the wave equation. The last additional term with the initial data is due to the presence of the boundary of the region.

3.6 Integral Representation of Dynamic Analogue of Gauss Formula ($N = 2$)

Using relation (3.32), the dynamic analogue of the Gauss formula can be written in integral form.

Lemma 3.4 *For $N = 2$ the dynamic analogue of Gauss formula has the following form:*

$$2\pi H_S^-(x)H(t) =$$
$$= V.P.\int_{S_\tau(x)} \frac{1}{\sqrt{1-(r/ct)^2}}\frac{\partial}{\partial n(y)}\left(\ln\frac{1}{r}\right)dS(y) + \frac{\partial}{c\partial t}\int_{S_t^-(x)} \frac{dV(y)}{\sqrt{c^2t^2-r^2}},$$

where the integral in the sense of the Principal Value is taken for the boundary points.

Proof. The formula of Lemma 6.1, taking into account (3.32)–(3.33), can be written as

$$2\pi H_S^-(x)H(t) = -\int\limits_{S_t(x)} \frac{ct}{\sqrt{c^2t^2-r^2}}\frac{1}{r}\frac{\partial r}{\partial n(y)}dS(y) + \frac{\partial}{c\partial t}\int\limits_{S_t^-(x)} \frac{dV(y)}{\sqrt{c^2t^2-r^2}} \quad (3.34)$$

We show that this equality, which is valid in the region of regularity ($x \notin S$), is also preserved for $x \in S$ if we take the singular integral on the left, which contains a strong singularity in r, in the sense of the principal value. Similarly (3.34) for a domain with a punctured $\varepsilon-$ neighborhood of x, we obtain

$$-\int\limits_{S_\varepsilon(x)+\Gamma_\varepsilon^-(x)} \frac{ct\,H(ct-r)}{\sqrt{c^2\tau^2-r^2}}\frac{1}{r}\frac{\partial r}{\partial n(y)}dS(y) + \frac{\partial}{c\partial t}\int\limits_{S_\varepsilon^-(x)} \frac{H(ct-r)dV(y)}{\sqrt{c^2\tau^2-r^2}} = 0,$$

$$-\int\limits_{S_\varepsilon(x)+\Gamma_\varepsilon^+(x)} \frac{ct\,H(ct-r)}{\sqrt{c^2\tau^2-r^2}}\frac{1}{r}\frac{\partial r}{\partial n(y)}dS(y) + \frac{\partial}{c\partial t}\int\limits_{S_\varepsilon^-(x)+Ш_\varepsilon(x)} \frac{H(ct-r)}{\sqrt{c^2\tau^2-r^2}}dV(y) = 2\pi.$$

On Γ_ε^+ $\frac{\partial r}{\partial n(y)} = 1$, on Γ_ε^- $\frac{\partial r}{\partial n(y)} = -1$, therefore, if both equalities are added and divided by 2, taking into account the symmetry properties of the integrands, we obtain

$$-\int\limits_{S_\varepsilon(x)} \frac{ct\,H(ct-r)}{\sqrt{c^2\tau^2-r^2}}\frac{1}{r}\frac{\partial r}{\partial n(y)}dS(y) +$$
$$+\frac{\partial}{c\partial t}\left(\int\limits_{S_\varepsilon^-(x)} \frac{H(ct-r)dV(y)}{\sqrt{c^2\tau^2-r^2}} + \int\limits_{Ш_\varepsilon(x)} \frac{H(ct-r)dV(y)}{\sqrt{c^2\tau^2-r^2}}\right) = \pi. \quad (3.35)$$

The last integral is calculated by passing to the polar coordinate system.
At $\varepsilon < ct$

$$I_\varepsilon(x) = \int\limits_{Ш_\varepsilon(x)} \frac{H(ct-r)dV(y)}{\sqrt{c^2\tau^2-r^2}} = \int\limits_0^{2\pi} d\theta \int\limits_0^{\varepsilon} \frac{rdr}{\sqrt{c^2t^2-r^2}} = 2\pi\left(ct - \sqrt{c^2t^2-\varepsilon^2}\right).$$

From here follow

$$\lim_{\varepsilon\to 0}\frac{\partial I_\varepsilon}{\partial t} = 2\pi c \lim_{\varepsilon\to 0}\left(1 - \frac{ct}{\sqrt{c^2t^2-\varepsilon^2}}\right) = 0 \text{ for } \forall t > 0. \quad (3.36)$$

Passing to the limit on $\varepsilon \to 0$ in equality (3.35) with (3.36) taken into account, for $x \in S$, $t > 0$ we obtain

$$-\text{V.P.} \int\limits_{S_t(x)} \frac{ct}{\sqrt{c^2t^2 - r^2}} \frac{1}{r} \frac{\partial r}{\partial n(y)} dS(y) + \frac{\partial}{c\partial t} \int\limits_{S_t^-(x)} \frac{dV(y)}{\sqrt{c^2t^2 - r^2}} = \pi.$$

That is, taking into account the definition of $H_S^-(x)$, formula (3.34) is valid for any x. From this formula, using elementary transformations, we obtain the formula of the lemma.

3.7 The Solution of Nonstationary Boundary Value Problems in 3D-Space: Singular Boundary Integral Equations

Let us construct the solution to the problem in three-dimensional space.

Theorem 3.8 *For $N = 3$ the solution of the initial-boundary value problem is representable in the form:*

for $x \in S^-$

$$4\pi u(x,t) =$$

$$= \int\limits_{S_t(x)} \left\{ \frac{1}{r} \frac{\partial u(y, t - r/c)}{\partial n(y)} - u\left(y, t - \frac{r}{c}\right) \frac{\partial}{\partial n(y)} \frac{1}{r} + c^{-1} \dot{u}\left(y, t - \frac{r}{c}\right) \frac{\partial \ln r}{\partial n(y)} \right\} dS(y) -$$

$$- \int\limits_{S_t^-(x)} r^{-1} G(x, t - r/c)\, dV(y) + c^{-1} \frac{\partial}{\partial t} \int\limits_{S_t(x)} u_0(y) \frac{\partial \ln r}{\partial n(y)} dS(y) +$$

$$+ \int\limits_{r=ct} \frac{\dot{u}_0(y)}{c^2 t} H_S^-(y) dS(y) + \frac{\partial}{\partial t} \int\limits_{r=ct} \frac{u_0(y)}{c^2 t} H_S^-(y) dS(y);$$

for $x \in S$

$$2\pi u(x,t) = \int\limits_{S_t(x)} \left\{ \frac{1}{r} \frac{\partial u(y, t - r/c)}{\partial n(y)} + \frac{\dot{u}(y, t - r/c)}{c} \frac{\partial \ln r}{\partial n(y)} \right\} dS(y) -$$

$$-\text{V.P.} \int\limits_{S_t(x)} u(y, t - r/c) \frac{\partial}{\partial n(y)} \frac{1}{r}\, dS(y) -$$

$$- \int\limits_{S_t^-(x)} \frac{G(x, t - r/c)}{r} dV(y) + c^{-1} \frac{\partial}{\partial t} \int\limits_{S_t(x)} u_0(y) \frac{\partial \ln r}{\partial n(y)} dS(y) +$$

$$+ \int\limits_{r=ct} \frac{\dot{u}_0(y)}{c^2 t} H_S^-(y) dS(y) + \frac{\partial}{\partial t} \int\limits_{r=ct} \frac{u_0(y)}{c^2 t} H_S^-(y) dS(y).$$

Proof. For $N = 3$, the fundamental solution of the wave equation—the singular generalized function—is a simple layer on the surface of the cone $K_t = \{(x, t) : \|x\| = ct\}$, which has the form:

$$\hat{U}(x,t) = -\frac{1}{4\pi\|x\|}\delta\left(t - \frac{\|x\|}{c}\right). \tag{3.37}$$

For any $\varphi(x, t) \in D(R^4)$, (3.37) defines a linear functional of the form (3.37):

$$\left(\hat{U}, \varphi\right) = -\frac{1}{4\pi}\int_{R^3} \frac{\varphi(x, \|x\|/c)}{\|x\|} dV(x).$$

By formulas (3.23) we find the corresponding antiderivatives:

$$\hat{W}(x,t) = -\frac{H(ct - \|x\|)}{4\pi\|x\|},$$

$$\hat{H}(x,m,t) = \frac{1}{4\pi}\frac{(x,m)}{\|x\|^2}\left(c^{-1}\delta\left(t - \frac{\|x\|}{c}\right) + \frac{H(ct - \|x\|)}{\|x\|}\right). \tag{3.38}$$

To construct integral representations of dynamic analogues of the Green and Gauss formulas, we use formulas (1.43)–(1.46).

We calculate the convolutions in the formula of Theorem 7.1 taking these relations into account. The first term is equal to

$$\hat{U}(x,t) * \frac{\partial u}{\partial n}\delta_S(x)H(t) = -\frac{H(t)}{4\pi}\int_{S_t(x)} \frac{1}{r}\frac{\partial u(y, t - r/c)}{\partial n(y)} dS(y).$$

Second term

$$-4\pi\ \hat{W}_{,j} * \dot{u}(x,t)n_j(x)\delta_S(x)H(t)$$

is equal to the sum of two convolutions:

$$-\frac{1}{4\pi c}\frac{x_j}{\|x\|^2}\delta(t - \|x\|/c) * \dot{u}(x,t)n_j(x)\delta_S(x)H(t) =$$

$$= \frac{H(t)}{c}\int_{S_t(x)} r^{-1}\dot{u}(y, t - r/c)r_{,j}n_j(y)dS(y) = c^{-1}H(t)\int_{S_t(x)} \dot{u}(y, t - r/c)\frac{\partial \ln r}{\partial n(y)} dS(y),$$

$$-\frac{x_j}{R^3}H(ct-\|x\|)*\dot{u}(x,t)n_j(x)\delta_S(x)H(t)=H(t)\int_0^t d\tau\int_{S_t(x)}\frac{r_{,j}n_j(y)}{r^2}\dot{u}(y,t-\tau)\,dS(y)=$$

$$=H(t)\int_{S_t(x}\frac{\partial r}{r^2\partial n(y)}dS(y)\int_0^t\dot{u}(y,t-\tau)d\tau\ =$$

$$=H(t)\int_{S_t(x)}u_0(y)\frac{\partial r^{-1}}{\partial n(y)}dS(y)-H(t)\int_{S_t(x)}u(y,t)\frac{\partial}{\partial n(y)}\frac{1}{r}\,dS(y).$$

Third term

$$-4\pi\,\hat{W}_{,j}\underset{x}{*}u_0(x)n_j(x)\delta_S(x)=$$

$$=-\frac{x_j}{\|x\|^2}\left(c^{-1}\delta(t-R/c)+\frac{H(ct-\|x\|)}{\|x\|}\right)\underset{x}{*}u_0(x)n_j(x)\delta_S(x)=$$

$$=\frac{\partial}{c\partial t}\int_{S_t(x)}\frac{y_j-x_j}{r^2}u_0(y)n_j(y)dS(y)+\int_{S_t(x)}\frac{y_j-x_j}{r^3}u_0(y)n_j(y)dS(y)=$$

$$=\frac{\partial}{c\partial t}\int_{S_t(x)}\frac{\partial\ln r}{\partial n(y)}u_0(y)dS(y)-\int_{S_t(x)}\frac{\partial r^{-1}}{\partial n(y)}u_0(y)dS(y).$$

The last three terms give the well-known Kirchhoff formula for solving the Cauchy problem (p. 226, [3]). Adding these convolutions, we obtain the formula of the theorem. In the part that depends on the initial data, the term due to the presence of the boundary S is added to the formula of the theorem. It vanishes at $t>t^*(x)$, $t^*(x)=\max\limits_{y\in S}\frac{\|x-y\|}{c}$ since $S_t(x)=S$ and the integral does not depend on t.

In the first formula of the theorem for $x\notin S$, $t>0$ all integrals exist. The proof of it when $x\in S$ similar to paragraph 1.5. Moreover, the second term on the right has a strong feature, since in this case, on Γ_ε^-

$$\lim_{\varepsilon\to0}\int_{\Gamma_\varepsilon^-}u\Big(y,t-\frac{r}{c}\Big)\frac{\partial}{\partial n(y)}\frac{1}{r}dS(y)=\lim_{\varepsilon\to0}\frac{1}{\varepsilon^2}\int_{\Gamma_\varepsilon^-}u\Big(y,t-\frac{r}{c}\Big)dS(y)=-2\pi u(x,t).$$

It is clear that for $x\in S$ the formula of Theorem 3.8 preserves the form if the corresponding singular integral is understood in the sense of the principal value $H_S^-(x)$ and the value on the boundary S.

For $x\in S$, the formula gives singular BIEs for solving the second boundary value problem. For the first boundary value problem, the same formula gives a BIE, but not a singular one. These equations are also not Fredholm.

The solution to the problem of their solvability, as in the flat case, is offered to the interested reader.

Using relations (3.30) and (3.38), the dynamic analogue of the Gauss formula can also be written in integral form. We state the result.

Lemma 3.5 *For $N = 3$ the dynamic analogue of the Gauss formula for the wave equation has the following integral representation:*

$$4\pi H_S^-(x)H(t) = \int\limits_{S_t(x)} \frac{\partial r^{-1}}{\partial n(y)} dS(y) + \frac{1}{c}\frac{\partial}{\partial t}\left\{ \int\limits_{r=ct} \frac{H_S^-(y)}{r} dS(y) + \int\limits_{S_t(x)} \frac{\partial \ln r}{\partial n(y)} dS(y) \right\}.$$

At $t > t^*(x)$ from here the well-known Gauss formula follows (p. 405, [3]):

$$\int\limits_S \frac{\partial r^{-1}}{\partial n(y)} dS(y) = 4\pi H_S^-(x). \tag{3.39}$$

Proof of this formula is similar to the proof in Lemma 6.1. However, note that $N = 2$ from the formulas of Lemma 6.1 does not follow a two-dimensional analogue of the Gauss formula.

3.8 The Solution of Nonstationary Boundary Value Problems for D'Alembert Equation ($N = 1$)

In the spatially one-dimensional case, the domain of the solution in R^1 is the segment $S^- = [a_1, a_2]$ of length $d = a_2 - a_1$, and the boundary is its edges $S = \{x = a_1, x = a_2\}$; the characteristic function of the domain of the solution $H_D^-(x, t) = H_S^-(x)H(t), H_S^-(x) = H(a_2 - x)H(x - a_1)$. The solution of this problem is given by the following theorem.

Theorem 3.9 *The solution of initial-boundary value problems for the d'Alembert equation for $N = 1$ has the following integral representation:*

$$2\hat{u}(x,t) = cH(ct - |x - a_2|) \int\limits_{\frac{|x-a_2|}{c}}^{t} u_{,x}(a_2, \tau) d\tau -$$

$$-cH(ct - |x - a_1|) \int\limits_{\frac{|x-a_1|}{c}}^{t} u_{,x}(a_1, \tau) d\tau +$$

$$+\operatorname{sgn}(x - a_1) H(ct - |x - a_1|) u\left(a_1, t - \frac{|x - a_1|}{c}\right) - \tag{3.40}$$

$$-\operatorname{sgn}(x - a_2) H(ct - |x - a_2|) u\left(a_2, t - \frac{|x - a_2|}{c}\right) +$$

$$+c^{-1} \int\limits_{a_1}^{a_2} \dot{u}_0(y) H(ct - |x - y|) dy + u_0(x + ct) H_S^-(x + ct) +$$

$$+u_0(x - ct) H_S^-(x - ct).$$

Proof. In the spatially one-dimensional case, the fundamental solution to the d'Alembert equation is the Riemann function (p. 206 [3]):

$$\hat{U}(x,t) = -\frac{c}{2} H(ct - |x|), \quad \hat{U}_{,t} = -\frac{c}{2} \delta(t - |x|/c) \tag{3.41}$$

$$\hat{W}(x,t) = -\frac{c}{2} H(ct - |x|)(ct - |x|), \quad \hat{W}_{,x}(x,t) = \frac{c}{2} H(ct - |x|) \operatorname{sgn} x, \tag{3.42}$$

where

$$\operatorname{sgn} x = \begin{cases} 1, & x > 0, \\ 0, & x = 0, \\ -1, & x < 0. \end{cases} \tag{3.43}$$

In the one-dimensional case, it is impossible directly to use formula (3.24) to construct an integral analogue of Green's formula, since some of its functions are not defined. A similar formula can be obtained by adding u to zero outside a given region and differentiating in the class of generalized functions. We act differently to use the formula (3.24).

We expand the domain of definition $u(x,t)$ to a band in $R^2 \times R^+$, $S^- = \{a_1 \le x_1 \le a_2, -\infty < x_2 < \infty\}$, $t > 0$. Then the boundary S will consist of two straight lines $x_1 = a_1$, $x_1 = a_2$, whose external normals have coordinates (-1.0) and (1.0), respectively, $\frac{\partial u}{\partial n} = \mp n \frac{\partial u}{\partial x_1}$ for $x \in S$, $H_S^-(x) = H(x_1 - a_1) H(a_2 - x_1)$, $n_1 \delta_S(x) = \frac{\partial H_S^-}{\partial x_1} = -\delta(x_1 - a_1) + \delta(x_1 - a_2)$. Based on the descent method, the formula of Theorem 3.4 transforms to the next

$$
\begin{aligned}
\hat{u} = \hat{U}_2 * \frac{\partial u}{\partial x} H(t)(\delta(x-a_2) - \delta(x-a_1)) + \\
+\frac{\partial \hat{W}_2}{\partial x} * \dot{u}(x,t) H(t)(\delta(x-a_2) - \delta(x-a_1)) + \\
+\frac{\partial \hat{W}_2}{\partial x} \underset{x}{*} u_0(x)(\delta(x-a_2) - \delta(x-a_1)) + c^{-2}\hat{U}_2 \underset{x}{*} \dot{u}_0(x) H_S^-(x) \\
+ + c^{-2}\left(\hat{U}_{2,t} \underset{x}{*} u_0(x) H_S^-(x)\right) + c^{-2}\hat{U}_2 * \hat{G}.
\end{aligned}
\tag{3.44}
$$

Here, all convolutions are already taken on the space $R^1 \times R^+$ or in R^1 with the Green function and its antiderivative at $N = 2$. Collapsing in x_2, we get

$$
\begin{aligned}
\hat{u}(x,t) = u(x,t) H(x-a_1) H(a_2 - x) H(t) = \\
= \hat{U}(x-a_2,t) \underset{t}{*} \frac{\partial u(a_2,t)}{\partial x} H(t) - \hat{U}(x-a_1,t) \underset{t}{*} \frac{\partial u(a_1,t)}{\partial x} H(t) + \\
+\hat{W}_{,x}(x-a_2,t) \underset{t}{*} \dot{u}(a_2,t) H(t) - \hat{W}_{,x}(x-a_1,t) \underset{t}{*} \dot{u}(a_1,t) H(t) + \\
+\hat{W}_{,x}(x-a_2,t) u_0(a_2) - \hat{W}_{,x}(x-a_1,t) u_0(a_1) - \\
-c^{-2}\left(\hat{U} \underset{x}{*} \dot{u}_0(x) H_S^-(x)\right) - c^{-2}\hat{U}_{,t} \underset{x}{*} u_0(x) H_S^-(x) + \hat{U} * \hat{G}.
\end{aligned}
\tag{3.45}
$$

We substitute (3.40)–(3.42) into (3.45) and perform the integration; as a result we obtain the formula of the theorem. It expresses $u(x,t)$ inside the region in terms of the initial and boundary values of the function itself and its first.

We show that it is also valid for $x = a_1,\ x = a_2$ (taking into account the definition of $H(0) = 0.5$. To do this, we write formula (3.24) for the intervals $\left(a_1', a_2\right) = (a_1 + \varepsilon, a_2)$ and $(a_1, a_2 - \varepsilon)$. Assuming $x = a_1$ we make the limit transition by $\varepsilon \to +0$:

$$
\begin{aligned}
0 = \lim_{\varepsilon \to 0} c \left\{ H(ct-d) \int_{\frac{|x-a_2|}{c}}^{t} u_{,x}(a_2,\tau) d\tau - H(ct-\varepsilon) \int_{\frac{\varepsilon}{c}}^{t} u_{,x}(a_1,\tau) d\tau \right\} + \\
+H(ct-d) u(a_2, t-d/c) + \operatorname{sgn}(-\varepsilon) H(ct) u(a_1,t) + c^{-1} \int_{a_1+\varepsilon}^{a_2} \dot{u}_0(y) H(ct - |a_1 - y|) dy + \\
+u_0(a_1 + ct) H_S^-(a_1 + ct) + u_0(a_1 - ct) H_S^-(a_1 - ct) = \\
= c \left\{ H(t - d/c) \int_{\frac{|x-a_2|}{c}}^{t} u_{,x}(a_2,\tau) d\tau - \int_{0}^{t} u_{,x}(a_1,\tau) d\tau \right\} + \\
+H\left(t - \frac{|a_1 - a_2|}{c}\right) u\left(a_2, t - \frac{|a_1 - a_2|}{c}\right) - H(t) u(a_1,t) + \\
+c^{-1} \int_{a_1}^{a_2} \dot{u}_0(y) H(t - |a_1 - y|/c) dy + u_0(a_1 + ct) H_S^-(a_1 + ct) + u_0(a_1 - ct) H_S^-(a_1 - ct).
\end{aligned}
$$

Transferring the term $-H(t)u(a_1,t)$ to the left side, taking into account the value of the characteristic function on the boundary, we obtain the formula of the theorem for the left boundary. Similar reasoning is carried out for $x = a_2$. As a result, at the ends of the segment (a_1,a_2) we have the following relations for determining unknowns:

for $x = a_1$

$$u(a_1,t) = c\left\{H(ct-d)\int_{\frac{d}{c}}^{t} u_{,x}(a_2,\tau)d\tau - H(t)\int_{0}^{t} u_{,x}(a_1,\tau)d\tau\right\}+$$

$$+H(ct-d)u\left(a_2, t-\frac{d}{c}\right)+c^{-1}\int_{a_1}^{a_2}\dot{u}_0(y)H(ct-|a_1-y|)dy+u_0(a_1+ct)H(d-ct)H(t);$$

for $x = a_2$

$$u(a_2,t) = c\left\{H(t)\int_{0}^{t} u_{,x}(a_2,\tau)d\tau - H(ct-d)\int_{\frac{d}{c}}^{t} u_{,x}(a_1,\tau)d\tau\right\}+$$

$$+c^{-1}\int_{a_1}^{a_2}\dot{u}_0(y)H(ct-|a_2-y|)dy+u_0(a_2-ct)H(d-ct)H(ct)+H(ct-d)u\left(a_1, t-\frac{d}{c}\right).$$

For given $u_{,x}(a_k,t)$, $k = 1, 2$, we obtain two functional equations with a delayed argument for determining u at the boundary of the region, which can be solved stepwise in time, starting from $t = 0$. With known $u(a_1,t)$, $u(a_2,t)$ is already a system of integral equations, where the desired functions $u_{,x}(a_k,t)$ ($k = 1, 2$) are under the sign of the integral. We can consider the mixed problem when a function is known at one end and its derivative at the other. In this case, this system of equations is resolving.

The study of the solvability of the constructed BIEs, which are singular for problems with boundary conditions of the Neumann type, is an independent problem of functional analysis, since the constructed equations do not belong to well-studied classical ones. However, we note that the use of computational methods based on the methods of the boundary element with the transition to discrete analogs of the BIE makes it possible to construct rather effectively solutions of such BIE.

References

1. Petrovsky, I.G.: Partial differential equations—Moskow "FIZMATGIZ", (1972)
2. Hadamard, J.: Le probleme de Cauchy et les equations aux derivees partielles lineaires hyperboliques. Paris, (1932)
3. Vladimirov, V.S.: Equations of mathematical physics—Moskow "Nauka", (1981)

Chapter 4
Boundary Integral Equations of Nonstationary Boundary Value Problems For Klein–Gordon–Fock Equation

Here we consider nonstationary boundary value problems for the Klein–Gordon–Fock equation with Dirichlet or Neumann conditions on the boundary of the domain of definition, and we prove the uniqueness of the formulated boundary value problems taking into account shock waves. Based on the method of generalized functions, a method of boundary integral equations is developed to solve the problems posed. Dynamic analogues of Green's formulas for their solutions in the space of generalized functions are obtained, and regular integral representations are constructed in the plane and three-dimensional cases. The resolving singular boundary integral equations are obtained for solving the stated initial-boundary value problems.

4.1 Klein–Gordon–Fock Equations: Shock Waves

The Klein–Gordon–Fock equation is considered:

$$\Box_c u + q(x)u = f(x,t), x \in R^N, t \in [0,\infty) \tag{4.1}$$

where the function *q(x)* is called *the scattering potential.* Equation (4.1) is of hyperbolic type. Its characteristic equation has the form:

$$\nu_t^2 - c^2\|\nu\|_N^3 = 0, \qquad \|\nu\|_N = \sqrt{\nu_j\nu_j}, \tag{4.2}$$

where $\nu(x,t) = (\nu_1, \ldots, \nu_{N,} \nu_t)$ is a normal vector to characteristic surface F in R^{N+1}. A cone of characteristic normals corresponds to it is a light cone for which $\nu_t < 0$. On F, the solution of Eq. (4.1) and its derivatives can be discontinuous. In R^N of the characteristic surface F there corresponds a wave front F_t (cross section F for a fixed t), which moves at a speed c:

L. Alexeyeva and A. Bayegizova, *Generalized Functions Method in Boundary Value Problems for Wave Equations*, Industrial and Applied Mathematics,
https://doi.org/10.1007/978-981-95-2802-8_4

$$c = -\frac{\nu_t}{\|\nu\|_N} \tag{4.3}$$

(hereinafter, everywhere to shorten the notation for repeated indices in the derivative, summation from 1 to N is carried out like a tensor convolution). Such solutions describe *shock waves.* Since (4.2) coincides with the characteristic equation of the classical d'Alembert wave equation, if the solution (4.1) is continuous:

$$[u(x,t)]_{F_t} = 0, \tag{4.4}$$

then at the wave fronts the Hadamard continuity conditions for jumps are satisfied: for $x \in F_t$

$$n_j[\dot{u}]_{F_t} = -c\big[u_{,j}\big]_{F_t}, \quad j = \overline{1,N} \tag{4.5}$$

$$[\dot{u}]_{F_t} = -cn_j\big[\,u_{,j}\big]_{F_t}, \tag{4.6}$$

where a wave vector $n(x,\ t)$ is a unit normal vector to F_t, directed toward the propagation of the wave front:

$$n_i = \nu_i/\|\nu\|_N, \quad i = \overline{1,N}. \tag{4.7}$$

Condition (4.5) is a consequence of the continuity condition (4.4) and ensures continuity on F_t of the tangent derivatives u. Condition (4.5) is the law of conservation of momentum at the fronts of shock waves. If in front of the wave front $u \equiv 0$, then at the wave front $(grad\ u, n) = -c^{-1}\dot{u}, \quad x \in F_t$. To study the solutions of the KGF Eq. (4.1), we use the GFM. To do this, we consider (4.1) on the space of generalized functions $D'(R^{N+1})$. Further, we will mark the ordinary locally integrable function with a cap if we consider it as a generalized.

If u has a finite discontinuity on F, then its generalized derivative

$$\hat{u}_{,j} = u_{,j} + [u]_F \nu_j \delta_F, \tag{4.8}$$

where the first term on the right is the classical derivative, the second is a simple layer on the surface F, $\boldsymbol{\nu}$ is its normal, and $\|\boldsymbol{\nu}\| = 1$.

Definition. The solutions of Eq. (4.1), continuous together with derivatives up to the second order inclusive almost everywhere, with the exception of a finite or countable number of discontinuity surfaces—wave fronts that are sufficiently smooth almost everywhere on which the conditions for jumps (4.5), (4.6) are satisfied, are called *classical solutions.*

Lemma 1.1 *If* $u(x,t)$ *is the classical solution of* (4.1), *then the corresponding regular generalized function* $\hat{u}(x,t)$ *is a generalized solution* (4.1).

Proof Using (4.8), we can sequentially determine the second derivatives. Given these equalities and (4.3), we obtain

$$(\Box_c + q(x))\hat{u} = f(x,t) + \left\{c^{-1}[u,_t]_{F_t} + \left[n_j u,_j\right]_{F_t}\right\}\|\nu\|_N \delta_F +$$
$$+c^{-1}\partial_t\left\{\|\nu\|_N [u]_{F_t}\delta_F)\right\} + \partial_j\left\{\|\nu\|_N [u]_{F_t} n_j \delta_F\right\}.$$

By virtue of (4.4) and (4.6), the densities of simple and double layers here are equal to zero on the right.

4.2 Statement of Nonstationary Boundary Value Problems for Klein–Gordon–Fock Equation: Law of Energy Conservation

Let a solution $u(x,t)$ (4.1) be constructed for $t \geq 0$ in the domain $S^- \in R^N$ bounded by the surface S. Let $m(x)$ be the vector of the external normal to S. The initial conditions are known: for $t = 0$

$$u(x,0) = u_0(x)\, for\, x \in S^- + S, \tag{4.9}$$

$$u,_t(x,0) = V_0(x)\, for\, x \in S^-. \tag{4.10}$$

Here we also consider two boundary value problems corresponding to the Dirichlet and Neumann boundary conditions: for $x \in S$

$$BVP\ I\ \ u(x,t) = u_S(x,t), \tag{4.11}$$

$$BVP\ II\ \ \frac{\partial u}{\partial n} = p(x,t). \tag{4.12}$$

At the fronts of shock waves, the Hadamard conditions (4.5), (4.6) for jumps are satisfied. Note that shock waves always occur if the condition for matching the initial and boundary data on velocities is not satisfied:

$$\dot{u}_S(x,0) = V_0(x), \quad x \in S \tag{4.13}$$

which is typical for physical tasks. In this case, at the initial moment of time, a shock wave front forms at the boundary S, which propagates at a speed of c in S^-. To construct continuously differentiable solutions, this condition is necessary. We will not enter it here.

It is assumed that the functions $u_0(x) \in C(S^- + S)$, $V_0(x) \in L_1(S^- + S)$, $p(x,t) \in L_1(D)$ and $u_S(x,t)$ are Hölder on S: $\forall\beta,\ \ 0 < \beta \leq 1$, for $\forall x \in S,\ y \in S,\ t \geq 0$

$$|u_S(x,t) - u_S(y,t)| \le const\|x-y\|^{\beta}. \tag{4.14}$$

We introduce the functions E—the energy density of the u-field and the Lagrange L-function:

$$E = 0,5\left(u,_t^2 + c^2\sum_{j=1}^{N} u,_j^2\right), \qquad L = 0,5\left(u,_t^2 - c^2\sum_{j=1}^{N} u,_j^2\right).$$

Lemma 4.1 *If* $u(x,t)$ *is a classical solution of the KGF-Eq.* (4.1), *then at the fronts of the shock waves the following conditions for jumps are satisfied:*

$$[E]_{F_t} = -c\left[\dot{u}\frac{\partial u}{\partial n}\right]_{F_t}, \ [L(x,t)]_{F_t} = \left(\dot{u}^- + c\frac{\partial u^-}{\partial n}\right)[\dot{u}]_{F_t}.$$

The consequence of this lemma is the law of conservation of energy.

Theorem 4.1 (Law of energy conservation). *If u(x, t) is classical solution of a boundary value problem, then.*

$$\int_{S^-} (E(x,t) - E_0(x))dV(x) + 0,5c^2\int_{S^-} q(x)\left(u^2(x,t) - u_0^2(x)\right)dV(x) =$$

$$= -c^2\int_0^t dt\int_{D^-} f(x,t)u,_t\,dV(x) + c^2\int_0^t\int_S (\dot{u}_S(x,t)p(x,t))dS(x)dt.$$

The proof of these lemma and theorem is the same way as for the wave equation (see Chap. 1, p. 1.2).

From Theorem 4.2 follows the uniqueness of the solutions of the boundary value problems.

Theorem 4.2 *If* $q(x) \ge 0$,, *then the classical solution of the first (second) boundary value problem for the KGF-equation is unique.*

Proof Due to the linearity of the problem, it is sufficient to prove the uniqueness of the zero solution. In this case, the right-hand side is zero, and the initial and corresponding boundary conditions are also zero. Then it is easy to see that from Theorem 4.1 it follows:

$$\int_{S^-}\left\{E(x,t) + 0,5c^2q(x)u^2(x,t)\right\}dV(x) = 0.$$

Since both terms are non-negative, therefore $E = 0, \quad u = 0$. The theorem is proven.

4.3 Dynamic Analogue of Green's Formula with a Constant Scattering Potential

Let us consider the case when the scattering potential is a constant:

$$q(x) = \pm m^2. \tag{4.15}$$

To construct a solution to boundary value problems, we use the method of generalized functions. To do this, we introduce regular generalized functions:

$$\hat{u} = u(x,t)H_D^-(x,t), \quad \hat{f} = f(x,t)H_D^-(x,t), \tag{4.16}$$

where $u(x, t)$ is a classical solution of a boundary value problem, $H_D^-(x,t) \equiv H_S^-(x)H(t)$ is a characteristic function of the domain of the solution of the space–time cylinder D^-, $H_S^-(x)$ is the characteristic function of the set S^- is 0.5 at its boundary S, and $H(t)$ is Heaviside function. Its generalized derivatives are equal

$$\frac{\partial H_D^-}{\partial x_j} = -n_j\delta_S(x)H(t), \quad \frac{\partial H_D^-}{\partial t} = -n_j H_S^-(x)\delta(t).$$

Let us consider the action of the KGF operator on $\hat{u}$ in D'. Since $[u]_S = -u$, performing generalized differentiation, similar to that in Chap. 1, and using conditions on the fronts, boundary and initial conditions, we obtain an equation for a generalized solution in the space of generalized functions:

$$\begin{gathered}\Box_c\hat{u} \pm m^2\hat{u} = \hat{f}(x,t) - \frac{\partial u}{\partial n}\delta_S(x)H(t) - H(t)\big(un_j\delta_S(x)\big)_{,j} - \\ -c^{-2}H_S^-(x)u_0(x)\dot{\delta}(t) - c^{-2}H_S^-(x)\dot{u}_0(x)\delta(t).\end{gathered} \tag{4.17}$$

Note that the densities of simple and double layers here are determined by the boundary conditions, some of which (depending on the BVP being solved) are known, and given initial conditions.

The solution to Eq. (4.17) is the convolution of the right side of the equation with the Green function $\hat{U}(x,t)$ satisfying the equation:

$$\Box_c\hat{U} \pm m^2\hat{U} = \delta(x)\delta(t) \tag{4.18}$$

and radiation conditions

$$\hat{U} = 0 \text{ at } t < 0, \ \hat{U} = 0 \text{ at } \|x\| > ct.$$

We obtain the solution (4.18) in the form of the following convolution of generalized functions:

$$\hat{u} = u(x,t)H_S^-(x)H(t) = -\hat{U} * \frac{\partial u}{\partial n}\delta_S(x)H(t) - \left(\hat{U} * un_j\delta_S(x)H(t)\right)_{,j} -$$
$$-\frac{1}{c^2}\left(\hat{U} \underset{x}{*} H_S^-(x)u_0(x)\right)_{,t} - c^{-2}\hat{U} \underset{x}{*} H_S^-(x)\dot{u}_0(x) + \hat{f}(x,t) * \hat{U}(x,t), \tag{4.19}$$

where the symbol $\underset{x}{*}$ means that the convolution is taken only for x. Moreover, the solution is unique in the class of functions that can be convolution with U. From this it is easy to obtain a solution to the Cauchy problem (in the absence of S, $S^- = R^N$).

Theorem 4.3 *The generalized solution to the Cauchy problem has the form:*

$$\hat{u}(x,t) = -c^{-2}\hat{U} \underset{x}{*} \dot{u}_0 - c^{-2}\left(\hat{U} \underset{x}{*} u_0\right)_{,t} + \hat{f} * \hat{U}. \tag{4.20}$$

Lemma 4.2 *For zero initial data and* $f = 0$, *the generalized solution has the form:*

$$\hat{u} = -\hat{U} * \frac{\partial u}{\partial n}\delta_S(x)H(t) - \hat{U}_{,j} * un_j\delta_S(x)H(t). \tag{4.21}$$

Formulas (4.19), (4.21) express the solution of BVPs in terms of the boundary values of the desired function and its derivative normal to the boundary, i.e., they are similar to the Green formula for solutions of the Laplace equation. However, due to the peculiarities of the fundamental solutions of hyperbolic equations at the wave front, the form of which depends on the dimension of space, their integral representation gives diverging integrals containing derivatives of the fundamental solution. To construct a regular integral representation, we regularize the formulas in the space of generalized functions. To do this, we introduce primitive functions with respect to t:

$$\hat{W} = \hat{U} * \delta(x)H(t) = \hat{U} \underset{t}{*} H(t),\ \partial_t\hat{W} = \hat{U}; \tag{4.22}$$

$$\hat{H}(x,n,t) = \frac{\partial \hat{W}}{\partial x_j}n_j = \frac{\partial \hat{W}}{\partial n}. \tag{4.23}$$

The functions $\hat{W}$ and $\hat{H}$ are also solutions of (4.1) for $\hat{f}(x,t) = H(t)\delta(x)$ and $\hat{f}(x,t) = H(t)\frac{\partial\delta(x,t)}{\partial n}$ respectively. The following theorem holds.

Theorem 4.4 *Generalized solution of boundary value problems satisfies the following equation in* $D'(R^{N+1})$:

$$\hat{u} = -\hat{U} * \frac{\partial u}{\partial n}\delta_S(x)H(t) - \hat{W}_{,j} * \dot{u}n_j(x)\delta_S(x)H(t) -$$
$$-\hat{W}_{,j} \underset{x}{*} u_0(x)n_j(x)\delta_S(x) - c^{-2}\hat{U} \underset{x}{*} H_S^-(x)\dot{u}_0(x) - \tag{4.24}$$
$$-c^{-2}\left(\hat{U} \underset{x}{*} H_S^-(x)u_0(x)\right)_{,t} + \hat{f} * \hat{U}.$$

Proof Since

$$\left(un_j\delta_S(x)H(t)\right)_{,t} = \dot{u}(x,t)n_j(x)\delta_S(x)H(t) + u(x,0)n_j(x)\delta_S(x)\delta(t),$$

using this equality and the property of convolution differentiation, we have

$$\left(\hat{U} * un_j\delta_S(x)H(t)\right)_{,j} = \left(\hat{W}_{,t} * un_j\delta_S(x)H(t)\right)_{,j} =$$
$$= \hat{W}_{,j} * \dot{u}(x,t)n_j(x)\delta_S(x)H(t) + \hat{W}_{,j} * u(x,0)n_j(x)\delta_S(x)\delta(t).$$

Using the delta function property,

$$\hat{W}_{,j} * u(x,0)n_j(x)\delta_S(x)\delta(t) = \hat{W}_{,j} \underset{x}{*} u_0(x)n_j(x)\delta_S(x).$$

Substituting these relations into (4.19), we obtain the formula of the theorem.

It follows from Theorem 4.3 that the solution to the problem is completely determined by the initial data, the boundary values of the normal derivative of the function $u(x,t)$, and its speed $\dot{u}(x,t)$. By analogy with the representation of solutions of the Laplace equation, we call these formulas *a dynamic analogue of the Green formula.*

Formula (4.24) allows you immediately to go to its integral notation without regularizing the integrands at the fronts.

Next, we consider the representation of solutions of boundary value problems for KGF equations in spaces of dimensions $N = 2$, 3, characteristic of mathematical physics problems. To avoid the cumbersomeness of formulas when constructing the integral representation of a dynamic analogue of Green's formula, we consider successively the solutions of two boundary value problems:

- Cauchy problem at $f(x,t) \neq 0$;
- boundary value problem with zero initial conditions and $f(x,t) = 0$.

Due to the linearity of the equations, the solution of the posed boundary value problems can be obtained by superposition of the solutions of these boundary value problems with the correction of the boundary conditions for the second problem, taking into account the boundary values of the solution to the Cauchy problem. A solution to the Cauchy problem for this equation was previously obtained by V. S. Vladimirov [1]. We will also get it here for a complete solution of the initial-boundary value problem in the notation accepted here.

4.4 Generalized Solution of Cauchy Problem for KGF Equation ($N = 2$)

Let us consider the Cauchy problem for a KGF equation of the form:

$$\Box_c \hat{u} \pm m^2 \hat{u} = \hat{f}(x,t) x \in R^N, \ t > 0 \tag{4.25}$$

where $\hat{f}(x,t)$, generally speaking, is a generalized function. We use the notation: $r = \|y - x\|$, $S_t^-(x) = \{y \in S^-, \ r < ct\}$, $S_t(x) = \{y \in S, \ \ r < ct\}$, which we will use further.

In a plane case N = 2. For $q(x) = -m^2$, the Green function of Eq. (4.25) is a regular function of the form:

$$\hat{U} = \frac{H(ct - \|x\|)}{2\pi} \frac{ch\left(m\sqrt{c^2t^2 - \|x\|^2}\right)}{\sqrt{c^2t^2 - \|x\|^2}} \tag{4.26}$$

with a weak feature at the front $\|x\| = ct$:

$$\hat{U} \sim \frac{1}{2\pi\sqrt{c^2t^2 - \|x\|^2}} \quad at \ \|x\| \to ct - 0. \tag{4.27}$$

Its carrier is a light cone $\|x\| \le ct$.

Theorem 4.5 *If the initial conditions are integrable:* $u_0(y) \in L_1(S^- + S)$, $\dot{u}_0(y) \in L_1(S^- + S)$, *then the solution of the Cauchy problem for Eq.* (4.25) *for* $q(x) = -m^2$, $N = 2$ *has the form:*

$$2\pi c^2 u(x,t) H(t) = \int\limits_{S_t^-(x)} \frac{ch\left(m\sqrt{c^2t^2 - r^2}\right)}{\sqrt{c^2t^2 - r^2}} \dot{u}_0(y) dV(y) +$$

$$+\partial_t \int\limits_{S_t^-(x)} \frac{ch\left(m\sqrt{c^2t^2 - r^2}\right)}{\sqrt{c^2t^2 - r^2}} u_0(y) dV(y)$$

$$-\int\limits_0^t d\tau \int\limits_{S_\tau^-(x)} \frac{ch\left(m\sqrt{c^2\tau^2 - r^2}\right)}{\sqrt{c^2\tau^2 - r^2}} f(y, t-\tau) dV(y).$$

Proof Integral notation of formula (4.24) leads to the formula of the theorem. All integrals are proper due to the regularity of integrands. The carrier of the kernel of integrals is a circle expanding over time with the center at the point x.

Note that if the initial functions and the right-hand side of Eq. (4.1) (source) belong to the class of singular functions admitting convolution with the Green's function of the equation, to construct a solution to the Cauchy problem, use the formula (4.24).

Similarly, we construct a solution to the Cauchy problem in case $q(x) = m^2$. The solution of the problem in this case allows analytic continuation. It can be obtained from the solution in Theorem 4.1 by replacing m with im.

4.5 Generalized Solution of Cauchy Problem for KGF Equation ($N = 3$)

For $N = 3$ the Green function of Eq. (4.1) with $q(x) = m^2$ is a singular generalized function of the form:

$$\hat{U} = \frac{H(t)}{4\pi r}\delta(ct - r) - \frac{mc}{4\pi} f_0(r, t), \tag{4.28}$$

where $r = \|x\|$, $H(t)\delta(ct - r)$ is a simple layer on a light cone $r = ct$, f_0 is defined by the expression:

$$f_0(r, t) = \frac{H(ct - r) J_1\left(m\sqrt{c^2t^2 - r^2}\right)}{\sqrt{c^2t^2 - r^2}} \tag{4.29}$$

$J_1(...)$ is a Bessel function. Because [2]

$$J_1(z) \sim 0.5z \; at \; z \to 0 \tag{4.30}$$

on the front $r = ct$, the second term has a finite jump:

$$\left[f_0(r, r/c)\right] = -\frac{m}{2}. \tag{4.31}$$

Theorem 4.6 *The solution of the Cauchy problem for the KGF Eq.* (4.1) *for* $N = 3$, $q(x) = m^2$ *has the form:*

$$\begin{aligned}
4\pi c u(x, t) = {} & (2ct)^{-1} \int\limits_{r=ct} \dot{u}_0(y) dS(y) - m \int\limits_{S_t^-(x)} \frac{J_1\left(m\sqrt{c^2t^2 - r^2}\right)}{\sqrt{c^2t^2 - r^2}} \dot{u}_0(y)\, dV(y) + \\
& + \frac{1}{2ct^2} \int\limits_{r=ct} u_0(y) dS(y) + \frac{1}{2ct} \partial_t \int\limits_{r=ct} u_0(y) dS(y) - \\
& - m\, \partial_t \left\{ \int\limits_{S_t^-(x)} \frac{J_1\left(m\sqrt{c^2t^2 - r^2}\right)}{\sqrt{c^2t^2 - r^2}} u_0(y)\, dV(y) \right\} - \\
& - mc^2 \int\limits_0^t d\tau \int\limits_{S_\tau^-(x)} \frac{f(y, t - \tau) J_1\left(m\sqrt{c^2\tau^2 - r^2}\right)}{\sqrt{c^2\tau^2 - r^2}} dV(y) + \\
& + \frac{c}{2} \int\limits_{S_t^-(x)} r^{-1} f(y, t - r/c)\, dV(y).
\end{aligned} \tag{4.32}$$

The proof follows from the presentation of a generalized solution for the Cauchy problem, taking into account the form of the fundamental solution (4.28). The solution of the Cauchy problem for Eq. (4.1) in the case of $q(x) = -m^2$ also admits analytic continuation by replacing m with im. It's form:

$$4\pi cu(x,t) = \frac{1}{2t}\int\limits_{r=ct} \dot{u}_0(y)dS(y) - m\int\limits_{S_t^-(x)} \frac{I_1\left(m\sqrt{c^2t^2-r^2}\right)}{\sqrt{c^2t^2-r^2}}\dot{u}_0(y)\,dV(y)+$$
$$+\frac{1}{2ct^2}\int\limits_{r=ct} u_0(y)dS(y) + \frac{1}{2ct}\partial_t\int\limits_{r=ct} u_0(y)dS(y)-$$
$$-m\partial_t\left\{\int\limits_{S_t^-(x)} \frac{I_1\left(m\sqrt{c^2t^2-r^2}\right)}{\sqrt{c^2t^2-r^2}}u_0(y)\,dV(y)\right\} + \frac{c}{2}\int\limits_{S_t^-(x)} r^{-1}f(y,t-\frac{r}{c})\,dV(y)-$$
$$-mc^2\int\limits_0^t d\tau\int\limits_{S_\tau^-(x)} \frac{f(y,t-\tau)I_1\left(m\sqrt{c^2\tau^2-r^2}\right)}{\sqrt{c^2\tau^2-r^2}}\,dV(y). \tag{4.33}$$

Here $I_1(...)$ is Bessel function of imaginary argument. If the initial functions and the right side of the equation (source) belong to the class of singular functions admitting convolution with the Green function, to construct the solution, one should use the convolutional formula (4.24). Below is an example for singular sources on the right side (4.26).

Example. *Cauchy problem for a dipole at the point* $x = 0$.
Given: $\hat{G} = bH(t)\sin\omega t\,\partial_1\delta(x)$, $u_0(x) = 0$, $\quad \dot{u}_0(x) = 0$.
We build its solution:

$$4\pi c^2 u = cb\sin\omega t\,H(t)\partial_1\delta(x) * H(t)\frac{\delta(ct-\|x\|)}{\|x\|} -$$
$$-mbH(t)\sin\omega t\,\partial_1\delta(x) * \frac{H(ct-\|x\|)J_1\left(\frac{m}{c}\sqrt{c^2t^2-\|x\|^2}\right)}{\sqrt{c^2t^2-\|x\|^2}} =$$
$$= cb\sin\omega t\,\partial_1\left\{H(t)\underset{t}{*}H(t)\frac{\delta(ct-\|x\|}{\|x\|})\right\} -$$
$$-mbH(t)\sin\omega t\,\partial_1\left\{H(t)\underset{t}{*}\frac{H(ct-\|x\|)J_1\left(\frac{m}{c}\sqrt{c^2t^2-\|x\|^2}\right)}{\sqrt{c^2t^2-\|x\|^2}}\right\}.$$

Compute all incoming convolutions in sequence:

$$H(t)\sin\omega t \underset{t}{*} H(t)\frac{\delta(ct-\|x\|)}{\|x\|} = \frac{H(ct-\|x\|)}{c\|x\|}\sin(\omega(t-\|x\|/c)),$$

$$\partial_1\delta(x) \underset{x}{*} \frac{H(ct-\|x\|)}{c\|x\|}\sin(\omega(t-\|x\|/c)) =$$

$$= -\frac{x_1 H(ct-\|x\|)}{c^2\|x\|^2}\left(\frac{c}{\|x\|}\sin(\omega(t-\|x\|/c)) + \omega\cos(\omega(t-\|x\|/c))\right),$$

$$H(t)\sin\omega t \underset{t}{*} \frac{H(ct-\|x\|)J_1\left(\frac{m}{c}\sqrt{c^2t^2-\|x\|^2}\right)}{\sqrt{c^2t^2-\|x\|^2}}$$

$$= H(ct-\|x\|)\int\limits_{\|x\|/c}^{t}\frac{\sin\omega(t-\tau)J_1\left(\frac{m}{c}\sqrt{c^2\tau^2-\|x\|^2}\right)}{\sqrt{c^2\tau^2-\|x\|^2}}d\tau.$$

As a result, we obtain the scalar field of the dipole in the form:

$$4\pi c^2 u(x,t)$$
$$= -\frac{bx_1 H(ct-\|x\|)}{\|x\|^2}\left(\frac{1}{\|x\|}\sin(\omega(t-\|x\|/c)) + \frac{\omega}{c}\cos(\omega(t-\|x\|/c))\right)$$
$$-mbH(ct-\|x\|)\left\{\int\limits_{\|x\|/c}^{t}\frac{\sin\omega(t-\tau)\partial_1 J_1\left(\frac{m}{c}z(x,\tau)\right)}{z(x,\tau)}d\tau\right\},$$

where $z(x,\tau) = \sqrt{c^2\tau^2-\|x\|^2}$.

At $z \to 0$

$$\partial_1 J_1\left(\frac{m}{c}\sqrt{c^2\tau^2-\|x\|^2}\right) = -\frac{m}{c}\left\{J_0\left(\frac{m}{c}z\right) - \frac{2c}{mz}J_1\left(\frac{m}{c}z\right)\right\}\frac{x_1}{z} = O(z^2)$$

since functions

$J_0\left(\frac{m}{c}z\right) = 1 + O(z^2)$, $J_1\left(\frac{m}{c}z\right) = \frac{m}{2c}z + O(z^3)$,

i.e., integrands in the integral have a removable singularity at the front and even are continuous on it.

Calculations of the dipole field for different values of the constants of the KGF equation and dipole parameters are presented in Figs. 4.1, 4.2, 4.3 and 4.4. Figure 4.1 shows the following parameter values: $\omega = 1$, $b = 1$, $c = 300$, $m = 1$ at fixed points in the environment. Notations P1(t) $= u(x,t)$ when $x =$ (0.1,0.1), P2(t) at (1,1), and P3(t) at (10,10) are introduced here. As you move away from the source, the oscillation amplitude decreases and tends to zero at $x \to \infty$. Figure 4.2 shows the dipole field $u(x,t)$ at the point $x =$ (0.1,0.1), at $b = 1$, $c = 300$, $m = 1$ for different frequencies ω. Designations: P1(t) $= u(x,t)$ at $\omega = 1$; P2(t) at $\omega = 3$.

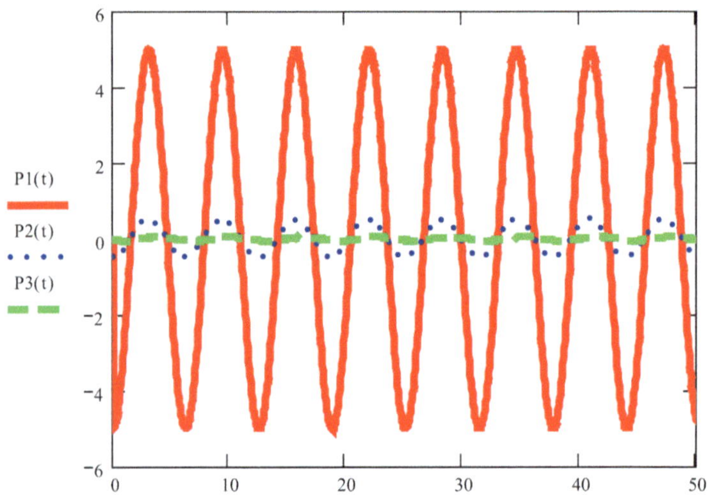

Fig. 4.1 Oscillograms of the field $u(x, t)$ for various x

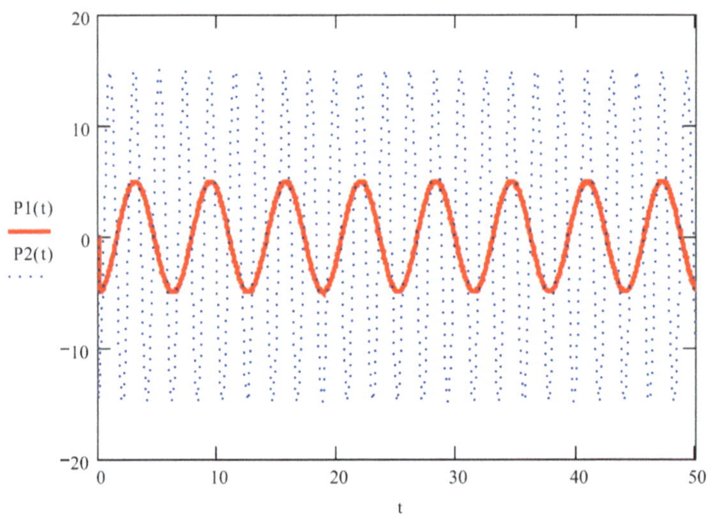

Fig. 4.2 Oscillograms of the dipole field $u(x, t)$ for various $\boldsymbol{\omega}$

An increase in frequency leads to a decrease in the period of oscillations $T = 2\pi/\omega$ f. Calculations showed that in this case, the field oscillogram in fixed currents of space is similar to a sinusoid, i.e., the main contribution is made by the first term.

Figure 4.3 shows $u(x, t)$: $\times\ 1 = 10$, $\times\ 2 = 10$, $b = 1$, $m = 1$, $\omega = 1$ for various values of the velocities c: P1(t) at $c = 300$; P2(t), $c = 3000$; P3(t), $c = 30{,}000$. The delay of the signal is clearly seen at low propagation velocities.

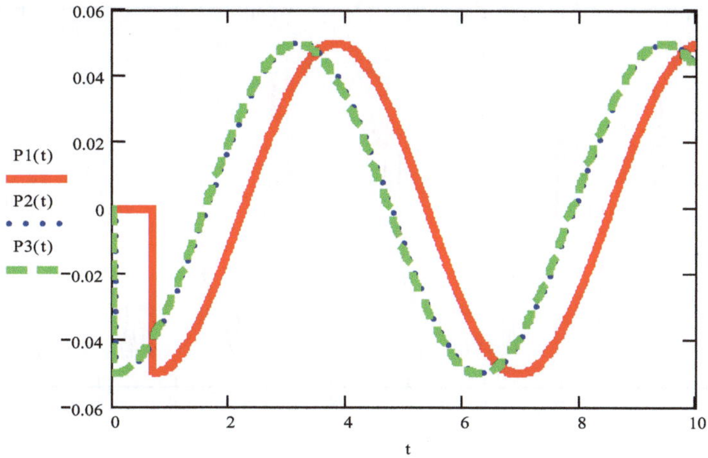

Fig. 4.3 $u(x, t)$ for various speeds c

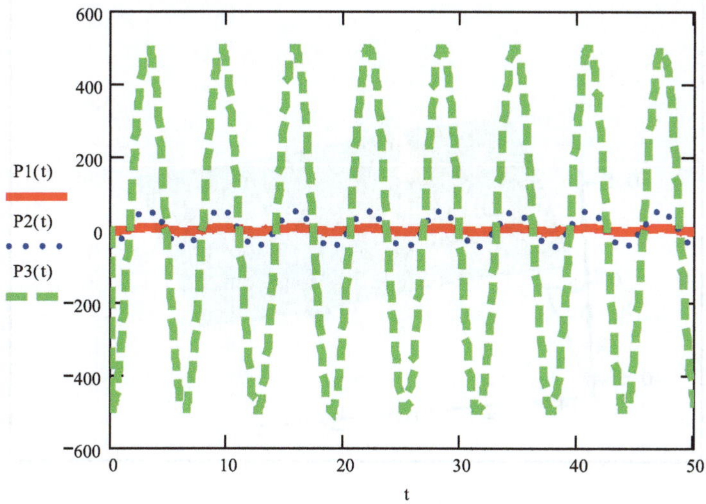

Fig. 4.4 Waveform $u(x, t)$ for various values of b

For small values of m, its change practically does not affect the nature of the oscillations. Figure 4.5 shows $u(x, t)$ ($\times 1 = 1$, $\times 2 = 1$, $b = 1$, $c = 10$, $\omega = 1$) for large values of m. As you can see, with such m, additional high-frequency oscillations appear on the waveform due to the resistance of the medium described by the additional term m^2u in the KGF Eq. (4.26).

The corresponding isolines of the dipole field are presented in Fig. 4.6. For the following parameter values: for $\omega = 1$, $b = 1$, $c = 300$, $m = 1$ at time $t = 1$ in the intervals (x1, x2) from -5 to 5.

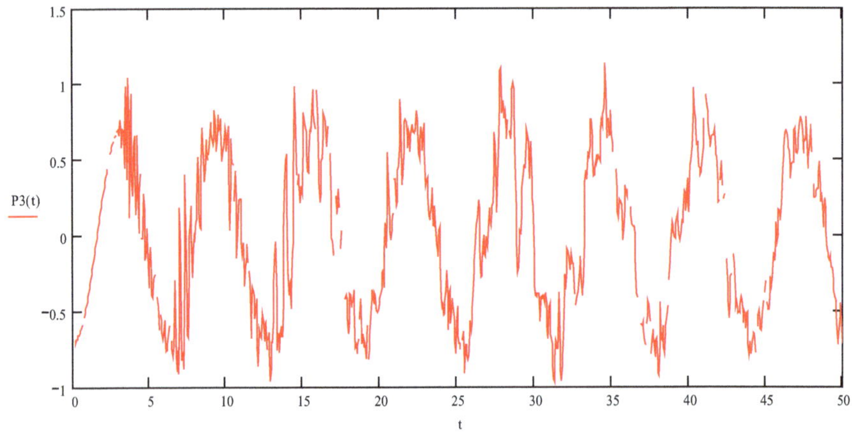

Fig. 4.5 Oscillogram $u(x, t)$ at $m = 50$

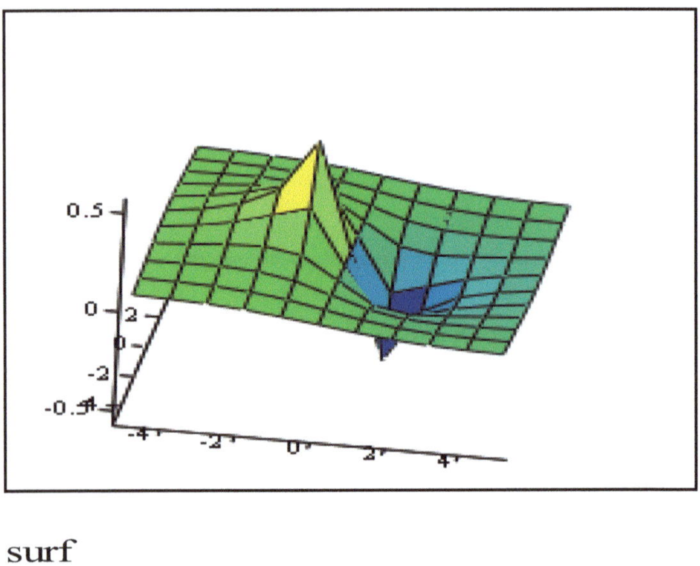

Fig. 4.6 Isolines of the dipole field $u(x, t)$: $t = 1$

4.6 Solutions and Singular Boundary Integral Equations of Plane Nonstationary Boundary Value Problems

Let us consider the solutions of the posed boundary value problems $q(x) = -m^2$ in the case N = 2. For the integral representation of the dynamic analogue of Green's formula, we calculate $\hat{W}$ and $\hat{H}$:

$$\hat{W} = \frac{1}{2\pi} d_0(r,t) * H(t) = \frac{1}{2\pi c} d_1(r,t), \tag{4.34}$$

$$\hat{H}(x,t,n) = \frac{1}{2\pi c} d_2(r,t)\frac{\partial r}{\partial n}, \quad r = \|x\|, \tag{4.35}$$

where

$$d_0(r,t) = H(ct - r)\frac{ch\left(m\sqrt{c^2t^2 - r^2}\right)}{\sqrt{c^2t^2 - r^2}},$$

$$d_1(r,t) = H(ct - r)\int_0^{\sqrt{c^2t^2-r^2}} \frac{ch(mz)}{\sqrt{z^2 + r^2}} dz,$$

$$d_2(r,t) = -H(ct - r)\left\{\frac{1}{r} - \frac{ch\left(m\sqrt{c^2t^2 - r^2}\right)}{ct}\right\}$$

$$-H(ct - r)r\int_0^{\sqrt{c^2t^2-r^2}} \frac{ch(mz)}{\left(\sqrt{z^2 + r^2}\right)^3} dz.$$

Let us consider the values of these functions at the front $\|x\| = ct$. From (4.34), (4.35) it follows that at the wave front

$$W|_{r=ct} = 0, \quad H|_{r=ct} = 0. \tag{4.36}$$

Therefore, unlike U, these functions are continuous at the front. When $r \to 0$ we obtain the asymptotic representation:

$$U = \frac{ch(mct)}{2\pi ct} + O(r), \qquad H = -\frac{1}{2\pi cr}\frac{\partial r}{\partial n} + O(1). \tag{4.37}$$

Now we turn to the integral notation of the dynamic analogue of the Gauss formula for $N = 2$.

Theorem 4.7 *The solution of the initial-boundary value problem for the KGF equation in the flat case, when* $q(x) = -m^2$, *can be represented as follows:*

for $x \notin S$

$$2\pi u(x,t)H(t) = \frac{H(t)}{c}\int\limits_{S_t(x)} H(ct-r)dS(y)\int\limits_{r/c}^{t}\left\{d_2(r,\tau)\frac{\partial r}{\partial n(y)}\dot{u}(y,t-\tau)\right\}d\tau$$

$$-\int\limits_{S_t(x)} H(ct-r)dS(y)\int\limits_{r/c}^{t}\{d_0(r,\tau)p(y,t-\tau)\}d\tau;$$

for $x \in S$

$$\pi\, cu(x,t)H(t) = \text{V.P.}\int\limits_{S_t(x)} H(ct-r)\frac{\partial r}{\partial n(y)}dS(y)\int\limits_{r/c}^{t} d_2(r,\tau)\dot{u}(y,t-\tau)d\tau -$$

$$-c\int\limits_{S_t(x)} H(ct-r)dS(y)\int\limits_{r/c}^{t} d_0(r,\tau)p(y,t-\tau)d\tau,\ r = \|y-x\|.$$

Proof In the flat case, the formula of Theorem 4.7, taking into account the carriers of the nuclei, can be written in the integral form:

$$2\pi\hat{u}(x,t)\int\limits_{S_t(x)} H(ct-r)dS(y)\int\limits_{r/c}^{t} H(x,y,\tau,n(y))\dot{u}(y,t-\tau)d\tau -$$

$$-H(t)\int\limits_{S_t(x)} H(ct-r)dS(y)\int\limits_{r/c}^{t} U(x,y,\tau)\frac{\partial u(y,t-\tau)}{\partial n(y)}d\tau.$$

Substituting the form of the nuclei, we obtain the first formula of the theorem. For $x \notin S$ all the integrals are convergent, since $r \neq 0$, and U has an integrable singularity at the front.

Let us prove the second formula for $x^* \in S$. We write an analog of Green's formula for a region with a punctured ε-neighborhood of a point x^*. We denote

$$O_\varepsilon(x) = \{y \in S : \|y-x^*\| < \varepsilon\},\quad \Gamma_\varepsilon^-(x) = \{y \in S^- : \|y-x^*\| = \varepsilon\},$$

$$\Omega_\varepsilon(x^*) = \left\{S_t(x^*) - O_\varepsilon(x^*)\right\} \cup \Gamma_\varepsilon^-(x^*).$$

Because x^* is outside the area bounded by the compound path $\Omega_\varepsilon(x^*)$, we have

$$\sum_{1\varepsilon}+\sum_{2\varepsilon}=\int\limits_{\Omega_\varepsilon(x^*)} H\left(ct-r\frac{\partial r}{\partial n(y)}\right)dS(y)\int\limits_{r*/c}^{t} d_2(r,\tau)\dot{u}(y,t-\tau)\frac{d\tau}{c}-$$

$$-\int\limits_{\Omega_\varepsilon(x^*)} H(ct-r)dS(y)\int\limits_{r/c}^{t} d_0(r,\tau)p(y,t-\tau)d\tau=0.$$

In this equality we pass to the limit for $\varepsilon\to+0$. On $\Gamma_\varepsilon^-(x^*)$ $r=\varepsilon$, arc length differential in polar coordinates: $dS(y)=\varepsilon d\varphi$. Therefore

$$\lim_{\varepsilon\to 0}\int\limits_{\Gamma_\varepsilon^-(x^*)} H(ct-\varepsilon)dS(y)\int\limits_{\varepsilon/c}^{t} d_0(\varepsilon,\tau)p(y,t-\tau)d\tau$$

$$=\lim_{\varepsilon\to 0}\varepsilon\pi\int\limits_{\varepsilon/c}^{t}\left\{\frac{ch(mc\tau)}{c\tau}p\left(x^*,t-\tau\right)\right\}d\tau$$

$$=\pi c\lim_{\varepsilon\to 0}\varepsilon\int\limits_{1}^{ct/\varepsilon}\left\{\frac{ch(m\varepsilon\tau)}{\tau}p\left(x^*,t-\frac{\varepsilon}{c}\tau\right)\right\}d\tau=0.$$

We have the last equality due to the inequality:

$$\left|\int\limits_{1}^{ct/\varepsilon}\left\{\frac{ch(m\varepsilon\tau)}{\tau}p\left(x^*,t-\frac{\varepsilon}{c}\tau\right)\right\}d\tau\right|\le ch(mct)\int\limits_{0}^{\infty}\left|p(x^*,\tau)\right|d\tau.$$

We consider the limit of the first integral:

$$\lim_{\varepsilon\to 0}\sum_{1\varepsilon}=\lim_{\varepsilon\to 0}\int\limits_{\Omega_\varepsilon(x^*)} H(ct-r)\frac{\partial r}{\partial n(y)}dS(y)\int\limits_{r^*/c}^{t} d_2(r,\tau)\dot{u}(y,t-\tau)\frac{d\tau}{c}$$

$$=V.P.\int\limits_{S_t(x^*)} H(ct-r)\frac{\partial r}{\partial n(y)}dS(y)\int\limits_{r^*/c}^{t} d_2(r,\tau)\dot{u}(y,t-\tau)\frac{d\tau}{c}-$$

$$-\lim_{\varepsilon\to 0}\int\limits_{\Gamma_\varepsilon^-(x^*)} H(ct-r)\frac{\partial r}{\partial n(y)}dS(y)\int\limits_{r/c}^{t} d_2(r,\tau)\dot{u}(y,t-\tau)\frac{d\tau}{c}$$

$$= V.P.\int\limits_{S_t(x^*)} H(ct-r)\frac{\partial r}{\partial n(y)}dS(y)\int\limits_{r^*/c}^{t} d_2(r,\tau)\dot{u}(y,t-\tau)\frac{d\tau}{c}$$

$$-\lim_{\varepsilon\to 0}\int\limits_{\Gamma_\varepsilon^-(x^*)} H(ct-r)\frac{\partial r}{\partial n(y)}dS(y)\int\limits_{r/c}^{t} d_2(r,\tau)\dot{u}(y,t-\tau)\frac{d\tau}{c} = I_S + -\lim_{\varepsilon\to 0} I_\varepsilon.$$

We calculate the last limit on the right side. Since on $\Gamma_\varepsilon^-(x^*)$ $\frac{\partial r}{\partial n} = -1$, hence $H = \frac{1}{2\pi c\varepsilon} + O(1)$, accordingly

$$\lim_{\varepsilon\to 0} I_\varepsilon = -\lim_{\varepsilon\to 0}\int\limits_{\Gamma_\varepsilon^-(x^*)} \frac{\partial r}{\partial n(y)}dS(y)\int\limits_{\varepsilon/c}^{t} d_2(\varepsilon,\tau)\dot{u}\big(x^*,t-\tau\big)\frac{d\tau}{c} =$$

$$= \lim_{\varepsilon\to 0}\pi\varepsilon\int\limits_{\varepsilon/c}^{t}\frac{\dot{u}(x^*,t-\tau)}{\varepsilon}d\tau = \pi\lim_{\varepsilon\to 0}\int\limits_{0}^{t}\dot{u}\big(x^*,t-\tau\big)\,d\tau = \pi\big(u\big(x^*,0\big)-u\big(x^*,t\big)\big)$$

$$= -\pi u\big(x^*,t\big).$$

Adding, taking into account the last equality, we obtain the second formula of the theorem. The theorem is proven.

In the case of the first BVP, the left-hand side of the equation of Theorem 6.1 and the first integral on the right are known, determined by the boundary conditions. Solving it, we determine the normal derivative of the desired function on the boundary, after which the formula of the theorem allows us to calculate the solution at any point in the domain of definition.

In the case of the second BVP, we have a BIE for determining the unknown boundary values of the desired function from the boundary values of its normal derivative. Solving it, we determine its values at the boundary, after which the solution of the boundary value problem is determined.

4.7 Dynamic Analogue of Green's Formula for Solutions of KGF Equations ($N = 3$)

Let us consider the case of $q(x) = m^2$ here. To construct a dynamic analogue of Green's formula in integral form, we also define $\hat{W}$ and $\hat{H}$ for $\hat{U}(x,t)$ (1.53). Calculating by formulas (4.23), (4.24), after some transformations, we obtain

$$4\pi c^2 \hat{W} = \frac{H(ct-r)}{r} - mf_1(r,t), \quad r = \|x\|,$$
$$\hat{H}(x,n,t) = -\frac{(x,n)}{2r}\left(\frac{\delta(ct-r)}{r} + \frac{H(ct-r)}{r^2} + 2mf_2(r,t)\right), \tag{4.38}$$

where

$$f_1(r,t) = H(ct-r)\int_0^{\sqrt{c^2t^2-r^2}} \frac{J_1(mz/c)}{\sqrt{z^2+r^2}}dz,$$

$$f_2(r,t) = rH(ct-r)\int_0^{\sqrt{c^2t^2-r^2}} \frac{J_1(mz/c)}{\left(\sqrt{z^2+r^2}\right)^3}dz + \frac{rH(ct-r)}{ct\sqrt{c^2t^2-r^2}}J_1\left(\frac{m}{c}\sqrt{c^2t^2-r^2}\right). \tag{4.39}$$

The values of these functions at the front $r = ct$: $f_1(r, r/c) = 0,\ f_2(r, r/c) = \frac{m}{2c}$. Consequently, W is continuous at the wave front; the last term in the representation H has a discontinuity of the first kind at the front.

When $r \to 0,\ t > 0$ the following asymptotics are true:

$$\hat{W}(x,t) = \frac{H(t)}{8\pi c^2 r} + O(1), \tag{4.40}$$

$$\hat{H}(x,n,t) = \frac{(e_x, n)}{8\pi c^2 r^2}H(t) + O(r^{-1}). \tag{4.41}$$

Next we use notations:

$$U(x,y,t) = \hat{U}(x-y,t),\ \ W(x,y,t) = \hat{W}(x-y,t),\ \ H(x,y,t,n) = \hat{H}(x-y,t,n).$$

Theorem 4.8 *A generalized solution of boundary value problems for a homogeneous KGF equation satisfying zero initial conditions can be represented as.*

$$4\pi\hat{u}(x,t) =$$

$$= -H(t)\int_{S_t(x)} \frac{1}{r}\frac{\partial u\left(y, t-\frac{r}{c}\right)}{\partial n(y)}dS(y) - c^{-1}H(t)\int_{S_t(x)} \dot{u}\left(y, t-\frac{r}{c}\right)\frac{(y-x, n(y))}{r^2}dS(y) +$$
$$+ mc^{-1}H(t)\int_{S_t^-(x)} f_0(r, r/c)\frac{\partial u(y, t-\tau)}{\partial n(y)}dV(y) +$$
$$+ mH(t)\int_0^t d\tau \int_{S_\tau(x)} f_2(r,\tau)\dot{u}(y, t-\tau)\ dS(y) + H(t)\int_{S_t(x)} u\left(y, t-\frac{r}{c}\right)\frac{(y-x, n(y))}{r^3}\ dS(y).$$

For $x \in S$ the last integral is singular, taken in the sense of the principal value.

Proof We write the case of a dynamic analogue of Green's formula (4.24) taking into account the form of the input functions. To do this, we compute convolution sequentially:

$$4\pi \hat{U}(x,t) * \frac{\partial u}{\partial n}\delta_S(x)H(t) =$$

$$= \int\limits_{S_t(x)} \frac{1}{r}\frac{\partial u(y,t-r/c)}{\partial n(y)}dS(y) - mc\int\limits_0^t d\tau \int\limits_{S_\tau(x)} f_0(r,\tau)\frac{\partial u(y,t-\tau)}{\partial n(y)}dS(y) =$$

$$= \int\limits_{S_t(x)} \frac{1}{r}\frac{\partial u(y,t-r/c)}{\partial n(y)}dS(y) - m\int\limits_{S_t^-(x)} f_0(r,r/c)\frac{\partial u(y,t-\tau)}{\partial n(y)}dV(y);$$

$$-4\pi\, \hat{W}_{,j} * \dot{u}\,(x,t)n_j(x)\delta_S(x)H(t) =$$

$$-\int\limits_{S_t(x)} \dot{u}(y,t-r/c)\,\frac{(y-x,n(y))}{cr^2}\,dS(y) -$$

$$-\int\limits_0^t d\tau \int\limits_{S_\tau(x)} \frac{(y-x,n(y))}{r^3}\,\dot{u}(y,t-\tau)\,dS(y)$$

$$+\int\limits_0^t d\tau \int\limits_{S_t(x)} \frac{(y-x,n(y))}{c^2r}f_2(r,\tau)\,\dot{u}(y,t-\tau)dS(y).$$

When $r = \|y-x\| \neq 0$ you can change the order of integration, therefore:

$$\int\limits_0^t d\tau \int\limits_{S_\tau(x)} \frac{(y-x,n(y))}{r^3}\,\dot{u}(y,t-\tau)\,dS(y) = -\int\limits_{S_t(x)} u(y,t-r/c)\,\frac{(y-x,n(y))}{r^3}\,dS(y).$$

Summing up, we obtain the formula of the theorem.

We note that for the points $x \in S$ with $t > 0$ all the kernels have no singularities, and the integrals on the right exist and define functions that are regular on this set. Since regular functions are left and right and they are equal on this set as generalized, by virtue of the Dubois-Reymond lemma, they are equal in the usual sense, like numerical functions.

Let $x^* \in S$. We write a dynamic analogue of Green's formula for a region with a punctured $\varepsilon-$ neighborhood of a point $x^* \in S,\ \varepsilon < ct$:

$$
\begin{aligned}
0 = &- \int\limits_{S_t(x)-O_\varepsilon(x)} \frac{1}{r} \frac{\partial u\left(y, t-\frac{r}{c}\right)}{\partial n(y)}\, dS(y) \\
&- \int\limits_{S_t(x)-O_\varepsilon(x)} \dot{u}\left(y, t-\frac{r}{c}\right) \frac{(y-x, n(y))}{cr^2}\, dS(y) + \\
&+ mc \int_0^t d\tau \int\limits_{S_\tau(x)-O_\varepsilon(x)} f_0(r,\tau) \frac{(y-x, n(y))}{r} \frac{\partial u(y, t-\tau)}{\partial n(y)}\, dS(y) + \\
&+ \int_0^t d\tau \int\limits_{S_\tau(x)-O_\varepsilon(x)} f_2(r,\tau) \dot{u}(y, t-\tau)\, dS(y) \\
&- \int\limits_{\Gamma_\varepsilon^-(x,t)} \frac{1}{r} \frac{\partial u(y, t-r/c)}{\partial n(y)}\, dS(y) - \\
&- \int\limits_{\Gamma_\varepsilon^-(x,t)} \dot{u}(y, t-r/c) \frac{(y-x, n(y))}{cr^2}\, dS(y) \\
&+ mc \int_0^t d\tau \int\limits_{\Gamma_\varepsilon^-(x,\tau)} f_0(r,\tau) \frac{(y-x, n(y))}{r} \frac{\partial u(y, t-\tau)}{\partial n(y)}\, dS(y) + \\
&+ \int_0^t d\tau \int\limits_{\Gamma_\varepsilon^-(x,\tau)} f_2(r,\tau)\, \dot{u}(y, t-\tau)\, dS(y) \\
&+ \int\limits_{S_t(x)-O_\varepsilon(x)} u(y, t-r/c) \frac{(y-x, n(y))}{r^3}\, dS(y) + \\
&+ \int\limits_{\Gamma_\varepsilon^-(x,t)} u(y, t-r/c) \frac{(y-x, n(y))}{r^3}\, dS(y).
\end{aligned}
$$

Now let's move on to the limit of $\varepsilon \to 0$. In the first integral, the integrand has a weak integrable singularity at $r = 0$. In the second integral, it does not have a singularity at $r = 0$. By virtue of this, the integrals over Γ_ε^- of these functions in the third and fourth terms tend to zero. It is easy to see that

$$
\begin{aligned}
&\lim_{\varepsilon \to 0} \int\limits_{S_t(x)-O_\varepsilon(x)} u(y, t-r/c) \frac{(y-x, n(y))}{2r^3}\, dS(y) = \\
&= V.P. \int\limits_{S_t(x)} u(y, t-r/c) \frac{(y-x^*, n(y))}{2r^3}\, dS(y).
\end{aligned}
$$

Moreover, the main value of the integral exists, since the integrand on a two-dimensional surface S has a singularity of order $1/r^2$, the function $u(x, t)$ is continuous on S, and the characteristic $(y - x^*, n(x^*))$ is antisymmetric at points opposite with respect to x^* (see [3]).

We consider the last limit. On $\Gamma_\varepsilon^-(x)$ we have $\|y - x\| = \varepsilon$, $n(y) = (x - y)/\varepsilon$,

$$\lim_{\varepsilon\to 0}\left\{\int\limits_{\Gamma_\varepsilon^-(x^*,t)} u(y, t - r/c)\frac{(y - x^*, n(y))}{r^3}dS(y)\right\} =$$

$$= \lim_{\varepsilon\to 0}\left\{\int\limits_{\Gamma_\varepsilon^-(x^*,t)} \big(u(y, t - r/c) - u(x^*, t)\big)\frac{(-n(y), n(y))}{\varepsilon^2}dS(y)\right\} -$$

$$-u(x^*, t)\lim_{\varepsilon\to 0}\left\{\int\limits_{\Gamma_\varepsilon^-(x^*,t)} \frac{dS(y)}{\varepsilon^2}\right\} = -2\pi u\big(x^*, t\big).$$

Passing to the limit in $\varepsilon \to 0$ and transferring the last term to the right side, we obtain the formula of the theorem. The theorem is proven.

For the first boundary value problem, the unknown normal derivative falls under the sign of the surface integral with a weakly polar core. The remaining terms are known. To solve the second initial-boundary value problem, the formula on the boundary gives a singular boundary integral equation. Note that, in addition to the boundary values of the function and its normal derivative, it includes the boundary values of the rate of change of this function, all with a time delay. This distinguishes a dynamic analogue of Green's formula from a similar formula for elliptic equations. In addition, the area of integration of integrands depends on time. This essentially distinguishes the boundary integral equations of hyperbolic boundary value problems from the BIE for elliptic problems. This is a new class of BIE in retarded potentials that requires special research using methods of functional analysis.

References

1. Vladimirov, V.S.: Generalized Functions in Mathematical Physics-Moskow "Nauka" (1979)
2. Special Functions Reference.: Abramowitz, M., Stigan, I. (eds.), Moskow "Nauka" (1979)
3. Muskhelishvili, N.I.: Singular integral equations-Moskow "Nauka", p 512 (1968)

Chapter 5
Generalized Solutions of Transport Boundary Value Problems for Wave Equation

Transport loads are very common in practice. By that we mean the loads moving in the medium, the form of which does not change with time, but their position in space changes. Dynamic deformation processes that occur in a medium under their influence propagate with velocities characterizing the elastic properties of the medium. For example, in an isotropic elastic medium, two *sound* velocities are associated with the propagation velocities of dilated and shear elastic waves (longitudinal and transverse waves). The ratio of the speed of the transport load to sound significantly affects the nature of the movement of the medium and the type of equations by which it is described.

Among the transport loads, we especially distinguish stationary ones that move in a fixed direction at a constant speed. This class of loads allows us to study diffraction processes in an analytical form at subsonic and supersonic speeds. For isotropic elastic media, this was shown, in particular, in [1–3], where, on the basis of the MGF, MBIE was developed for solving transport boundary value problems in elastic media with a cylindrical shape of the boundaries.

Here, we describe the main ideas of this method using the example of solving transport boundary value problems for the d'Alembert wave equation in cylindrical regions in the subsonic and supersonic cases.

5.1 Statement of Transport Boundary Value Problems

In the region $D^- \in R^{N+1}$ bounded by a cylindrical surface D whose generatrix is parallel to the X_{N+1} axis and the guide S is a smooth closed surface in R^N from the class of Lyapunov surfaces, we consider the class of solutions of the wave equation

$$\Delta_{N+1}u - c^{-2}\frac{d^2u}{dt^2} + g(\mathbf{x}, x_{N+1} + vt) = 0, \mathbf{x} = (x_1, ..., x_N), \tag{5.1}$$

L. Alexeyeva and A. Bayegizova, *Generalized Functions Method in Boundary Value Problems for Wave Equations*, Industrial and Applied Mathematics,
https://doi.org/10.1007/978-981-95-2802-8_5

describing the waves generated by a source moving along the axis of the cylinder with a speed v, $v > 0$, the direction of motion being opposite to the direction of the X_{N+1} axis. Hereinafter, Δ_{N+1} is the Laplace operator in R^{N+1}, $\|(\mathbf{x}, x_{N+1})\| = \left(\|\mathbf{x}\|^2 + x_{N+1}^2\right)^{1/2}$, $\|\mathbf{x}\| = (x_1^2 + \ldots x_N^2)^{1/2}$, $-\infty < t < \infty$. The action of the source is determined by the potential g(x,z), which is described by the function integrable on D^-:

$$g(\mathbf{x}, x_{N+1}) \to 0\, at\ \|(\mathbf{x}, x_{N+1})\| \to \infty. \tag{5.2}$$

If the domain D^- is bounded in x, the problem will be called *internal*, otherwise, *external*. Denote by $S^- = \{\mathbf{x}\colon (\mathbf{x}, x_{N+1}) \in D^-\}$ the section D^-. We consider two boundary value problems.

We suppose that there are Dirichlet or Neumann boundary conditions for $x \in S$:

BVP I

$$u = u_D(\mathbf{x}, x_{N+1} + vt),$$

$$u_D = O(|x_{N+1}|^{-1}), \quad \frac{\partial u_D}{\partial x_{N+1}} = o(|x_{N+1}|^{-1}) \quad \text{at } x_{N+1} \to \infty, \quad \forall t; \tag{5.3}$$

BVP II

$$\frac{du}{dn} = p_D(\mathbf{x}, x_{N+1} + vt), p_D = o(|x_{N+1}|^{-1})\text{at}\, x_{N+1} \to \infty, \forall t. \tag{5.4}$$

n is a unit vector of the outer normal to D: $n(\mathbf{x}, x_{N+1})$, $n_{N+1} = 0, \mathbf{x} \in S$.

It is required to construct a solution to these boundary value problems satisfying the damping condition at infinity:

$$u(x, t) \to 0\, \text{at} \|x\| \to \infty, \forall t. \tag{5.5}$$

To solve this problem, it is convenient to switch to a moving coordinate system $\mathbf{x}\prime = (x_1, \ldots, x_N, z)$, where $z = x_{N+1} + vt$. In the new system, the function $u = u(\mathbf{x}, z)$ is a solution to the equation

$$(1 - M^2)u_{,zz} + \Delta_N u + g(\mathbf{x}, z) = 0), \quad (x, z) \in D^- \tag{5.6}$$

where $M = v/c$ is *Mach number*. It is easy to reformulate the remaining conditions on u in a moving coordinate system.

The type of Eq. (5.6) depends on the Mach number; therefore, the solution to the problem is parametric in M. Three cases are possible depending on its value: *subsonic* ($M < 1$)—it's elliptical, *sound* ($M = 1$)—it's parabolic, and *supersonic* ($M > 1$)—it's hyperbolic, which determines the features and differences in the construction of solutions for each of them.

We use the method of generalized functions for this, which includes the following steps:

- statement of the boundary value problem in space $D'(R^{N+1})$;
- the construction of Green's function $U(\mathbf{x}, z)$ and other fundamental solutions of Eq. (5.1) in R^{N+1};
- construction of dynamic analogues of Green's formula and their regular integral representations;
- construction of dynamic analogs of the Gauss formula for fundamental solutions;
- construction of boundary integral equations that determine the solution to the problem.

At first we consider the subsonic case.

5.2 Uniqueness of the Solution of Transport Boundary Value Problem at Subsonic Speeds

At subsonic speeds $M < 1$, Eq. (5.6) is elliptical and has the form:

$$\Delta_N u + m^2 u,_{zz} + g(x, z) = 0, \quad (x, z) \in D^-, \tag{5.7}$$

Where $m = \sqrt{1 - M^2}$, $m > 0$. Here and further $x = \mathbf{x}$.

Lemma 5.1 *The classical solution of the first and second internal boundary value problems is unique.*

Proof By virtue of the linearity of the problem, it suffices to prove the uniqueness of the zero solution to the homogeneous boundary value problem.

Let $u(\mathbf{x},z)$ be a solution of Eq. (5.6) for $g = 0$ satisfying zero boundary conditions. Multiplying (5.6) by $u,_z$ and integrating over a finite cylinder $C^- = D^- \cap \{(x, z) : a < z < b\}$, where a, b are arbitrary ordered numbers, we obtain

$$\int_{S^-} dS(x) \int_a^b (0, 5m^2((u,_z)^2),_z + u,_z \Delta u]dz = 0.$$

It is convenient to transform equality using the first Green formula for the second term (c.338 [4]) and integrating the first over z. Since the normal at the ends of the cylinder has the coordinates $(0, \pm 1)$ as a result, we obtain

$$(0,5m^2-1)\int\limits_S \left[(u,_z^2(x,b)-u,_z^2(x,a))\right]dV(x)-\int\limits_C u,_z u,_i\, dV(x,z)+ \\ +\int\limits_C u,_z(x,z)\frac{\partial u}{\partial n(x)}dS(x,z)=0. \tag{5.8}$$

Here C is boundary surface of the cylinder C^-. Next, we transform the second integral of this equality to the form

$$\int\limits_{C^-} u,_{iz} u,_i\, dV(x,z)=0,5\int\limits_{C^-}\sum_{i=1}^{N}((u,_i(x,z))^2),_z\, dV(x,z)$$

$$=0,5\int\limits_{S^-}\sum_{i=1}^{N}((u,_i^2(x,b)-u,_i^2(x,a))dS(x).$$

The last surface integral in (5.8) is equal to zero due to the boundary conditions of the first or second boundary value problem. As a result, expression (5.8) is reduced to the form

$$\int\limits_{S^-}\left[(M^2+1)u,_z^2(x,b)+\sum_{i=1}^{N}u,_i^2(x,b)\right]dS(x) \\ =\int\limits_{S^-}\left[(M^2+1)u,_z^2(x,a)+\sum_{i=1}^{N}u,_i^2(x,a))\right]dS(x). \tag{5.9}$$

We pass here to the limit of $b\to+\infty$. Taking into account the decay conditions of the solution at infinity (5.6), we obtain the equality

$$\int\limits_{S^-}\left[(M^2+1)u,_z^2(x,a)+\sum_{i=1}^{N}u,_i^2(x,a))\right]dS(x)=0. \tag{5.10}$$

Since the integrand is non-negative, it is necessary that $u=const$. Due to damping conditions and the arbitrariness of $a,\ u\equiv 0$. The lemma is proven.

Lemma 5.2 *The classical solution of the first (second) external boundary value problem is unique if there exists $a>1$, so that*

$$|u,_z|\le O\big(\|x,z\|^{-(N+\alpha)/2}\big)npu\|x,z\|\to\infty. \tag{5.11}$$

Proof is similar to the above. The last condition of the lemma is necessary for the existence of integrals over an unbounded domain S^- and their tendency to zero in (5.8) for $b\to\infty$.

5.3 Green's Function in $D'(R^{N+1})$ at Subsonic Speeds:An Analogue of Gauss Formula

We denote Green's function by $\hat{U}(x, z)$. It is the solution of Eq. (5.6) in $D'(R^{N+1})$ for $g(x, z) = \delta(x)\delta(z)$, satisfying the damping conditions at infinity:

$$\Delta_m^+ \hat{U} + \delta(x, z) = 0,\ U(x, z) \to 0\, at\ \|x, z\| \to \infty. \tag{5.12}$$

Here and below we denote $\Delta_m^\pm = \Delta_N \pm m^2 \frac{d^2}{dz^2}$ according to the sign. It can be easily obtained from the well-known fundamental solution of the Laplace equation (p.208 [4]) by coordinate transformation $y = (x_1, \ldots, x_N, \frac{z}{m})$:

$$\hat{U}(x, z) = \frac{m^{N-2}}{\sigma_{N+1}(N-1)}(m^2\|x\|^2 + z^2)^{\frac{1-N}{2}}, \quad N \geq 2 \tag{5.13}$$

$$\hat{U}(x, z) = -\frac{1}{2\pi m}\ln\left[\frac{\sqrt{z^2 + m^2x^2}}{m}\right], \qquad N = 1. \tag{5.14}$$

Here σ_{N+1} is unit area in R^{N+1}:$\sigma_{N+1} = 2\pi^{\frac{N+1}{2}}\Gamma[(N+1)/2]$, where $\Gamma(z) = \int\limits_0^\infty t^{z-1}e^{-t}dt$ is gamma function. We introduce the derivative function

$$\hat{T}(x, z, \nu) = \frac{d\hat{U}}{dx_i}\nu_i + \frac{d\hat{U}}{dz}\nu_{N+1}, i = 1, \ldots, N, \tag{5.15}$$

which is also a solution (5.6) corresponding to $g = \frac{d\delta(x,z)}{d\nu}, \quad \nu = (\nu_1, ..., \nu_{N+1})$.

Calculating, we find.

for $N \geq 2$

$$\hat{T}(x, z, \nu) = -\frac{m^N}{\sigma_{N+1}}\left[\nu_{N+1}z + m^2\sum_{k=1}^{N}(\nu_k x_k)\right](z^2 + m^2\|x\|^2)^{-\frac{N+1}{2}}, \tag{5.16}$$

for $N = 1$

$$\hat{T}(x, z, \nu) = -\frac{1}{2\pi m}\frac{z\nu_2 + m^2x\nu_1}{z^2 + m^2x^2}, \qquad x = x_1. \tag{5.17}$$

The functions $\hat{U}, \hat{T}$ are homogeneous in (x, z) of degree (1-N) and (-N).

For $N \geq 2$

$$\hat{U}(x,z) = \frac{1}{\|(x,z)\|^{N-1}} U(x^o, z^o),\ \left\|(x^o, z^o)\right\| = 1,$$
$$\hat{T}(x,z,\nu) = \frac{1}{\|(x,z)\|^{N}} T(x^o, z^o, \nu), \tag{5.18}$$

possess the following symmetry properties:

$$\hat{U}(-x,-z) = \hat{U}(x,z) = \hat{U}(-x,z) = \hat{U}(x,-z),$$
$$\hat{T}(-x,-z,\nu) = -T(x,z,\nu) = T(x,z,-\nu). \tag{5.19}$$

The following lemma holds.

Lemma 5.3 *$\hat{T}(x,z,n)$ satisfies the Gauss formula:*

$$\int_{-\infty}^{+\infty} d\zeta \int_S T(y-x, \zeta - z, n(y))dS(y) + H_S^-(x) = \begin{cases} 0 & (external\quad BVP) \\ 1 & (internal\ BVP) \end{cases}.$$

For $(x,z) \notin D$ this integral is regular, and for $(x,z) \in D$ it is singular and taken in the Value Principle sense.

Here, the internal problem corresponds to a bounded, simply connected region of S^-, and to the external, a doubly connected unbounded region containing infinitely distant points.

Proof *(internal BVP)*. Let's fix $a > 0$. We collapse Eq. (5.12) with the characteristic function $H_S^-(x)H(a-|z|)$ of a bounded cylinder $C^- = \{(x,z) : x \in S, |z| \le a\}$. Using the convolution differentiation property, we obtain

$$-\hat{U}_{,i} * n_i(x)\delta_S(x)H(|z|-a) - m^2\hat{U}_{,z} * H_S^-(x)\ \text{sgn}\ z\ \delta(|z|-a)$$
$$+H_S^-(x)H(|a|-z) = 0.$$

Note that due to the finiteness of $H_S^-(x)H(|a|-z)$, all convolutions exist. For $x \notin S$, the equality can be rewritten in integral form, since $\hat{U}$ is regular GF and has a weak singularity at zero:

$$\int_{-a}^{+a} dt \int_S T(y-x, t-z, n(y,t))dS(y) +$$
$$+m^2 \int_{S^-} \left[\frac{dU(x-y, z-a)}{dz} - \frac{d\hat{U}(x-y, z+a)}{dz}\right] dV(y) + H_S^-(x)H(a-|z|) = 0. \tag{5.20}$$

At $a \to \infty$

$$\left| \int_{S^-} \frac{d\hat{U}(x-y, z \pm a)}{dz} dV(y) \right|$$

$$= \frac{m^N}{\sigma_{N+1}} \left| \int_{S^-} \frac{(z \pm a) dV(y)}{\left[(z \pm a)^2 + m^2 r^2\right]^{(N+1)/2}} \right| \le \frac{m^N}{\sigma_{N+1} |z \pm a|^N} \int_{S^-} dV(y) = \frac{const}{|z \pm a|} \to 0.$$

From (5.20) follows the formula of the lemma for $x \notin S$. We prove it for $x \in S$.

Let $x \in S$. We use the following notation: $D^+ = R_{N+1} - (D + D^-)$, $D_\varepsilon = D - O_\varepsilon$, $O_\varepsilon(x, z) = \{(y, t) \in D : \|(x - y, t - z)\| < \varepsilon\}$, $\Gamma_\varepsilon(x, z) = \{(y, t) : \|(x - y, t - z)\| = \varepsilon\}$, $Ш_\varepsilon^+(x,z) = Ш_\varepsilon(x,z) \quad D^+$,

$Ш_\varepsilon(x,z) = \{(y,t) : \|(x-y,t-z)\| < \varepsilon\}$, $Ш_\varepsilon^-(x,z) = Ш_\varepsilon(x,z) \cap D^-$. $\Gamma_\varepsilon^-, \Gamma_\varepsilon^+$ are the boundaries of $Ш_\varepsilon^-$ and $Ш_\varepsilon^+$ in D^- and D^+. A formula similar to (5.20) can be written for regions $D^- + Ш_\varepsilon^+(x,z)$ and $D^- - Ш_\varepsilon^-(x,z)$, $(x, z) \in D$:

$$\int_{D \backslash O_\varepsilon(x,z)} T(y - x, t - z, n(y, t)) dS(y, t) + \int_{\Gamma_\varepsilon^+(x,z)} T(y - x, t - z, \nu(y, t)) dS(y, t) -$$

$$-M^2 \int_{\Gamma_\varepsilon^+(x,z)} \frac{dU(x - y, z - t)}{dz} \nu_{N+1}(y, t) dS(y, t) + 1 = 0,$$

$$\int_{D \backslash O_\varepsilon(x,z)} T(y - x, t - z, n(y, t)) dS(y, t) + \int_{\Gamma_\varepsilon^-(x,z)} T(y - x, t - z, \nu(y, t)) dS(y, t) -$$

$$-M^2 \int_{\Gamma_\varepsilon^-(x,z)} \frac{dU(x - y, z - t)}{dz} \nu_{N+1}(y, t) dS(y, t) = 0.$$

Here $\nu(y,t)$ is a unit normal on a sphere centered at (x,z). On Γ_ε^- and Γ_ε^+: $\nu(y, \tau) = \nu(x - (y - x), z - (\tau - z))$. Therefore, by virtue of the antisymmetry of the function T and $U_{,z}$ (5.19), adding up these two equalities, dividing by 2, and passing to the limit by $\varepsilon \to \infty$, we obtain, taking into account the definition of $H_S^-(x)$, the formula of the lemma for boundary points.

Subtracting from these same relations, we obtain another useful equality, which we need in the future:

$$\lim_{\varepsilon \to 0} \int_{\Gamma_\varepsilon(x,z)} T\left[y - x, t - z, \frac{(y - x, t)}{\|(y - x, t)\|}\right] dS(y, t) + 1 = 0. \tag{5.21}$$

External BVP. In this case, we introduce an infinite cylinder C_ρ^- of sufficiently large radius ρ containing D. Similarly the above, for the region $C_\rho^- \cap D^-$ we obtain

$$\int_D T(y-x,t-z,n(y))dS(y,t)$$
$$+\int_{C_\rho} T\left[y-x,t-z,\frac{(y,0)}{\|y\|}\right]dS(y,t)+H_S^-(x)H(\rho-\|x\|)=0.$$

Since

$$\int_{C_\rho} T\left[y-x,t-z,\frac{(y,0)}{\|y\|}\right]dS(y,t)+H(\rho-\|x\|)=0,$$

therefore,

$$\int_D T(y-x,t-z,n(y))ds(y,t)+(H_S^-(x)-1)H(\rho-\|x\|)=0.$$

Passing to the limit with $\rho\to\infty$, we obtain the second formula of the lemma.

5.4 Dynamic Analogue of Green's Formula in $D'(R^{N+1})$

We introduce the characteristic function of the set D^-, defining it on the boundary of the region,

$$H_D^-(x,z)=H_S^-(x)1(z),\quad 1(z)\equiv 1,$$
$$\frac{dH_D^-(x,z)}{dx_j}=-n_j\delta_D(x,z)=-n_i\delta_S(x)1(z),$$

where $\delta_D(x,z)$ is a simple layer on D:

$$(\alpha(x,z)\delta_D(x,z),\varphi(x,z))=\int_D\alpha(x,z)\varphi(x,z)dD(x,z),\ \forall\varphi\in D(R_{N+1}),$$

and $\alpha(x,z)$ is the density of a simple layer which is a function integrable on D. Hereinafter $dS(\ldots)$ and $dV(\ldots)$ are differentials of surface area and volume according to the indicated integration set.

We introduce the function $\hat{u}(x,z)=u(x,z)H_D^-(x,z)$, where $u(x,z)$ is the classical solution of BVP under consideration in the domain $D+D^-$, now defined on the whole space R^{N+1}, and we will consider it as a regular generalized function from $D'(R^{N+1})$. Similarly $\hat{g}(x,z)=g(x,z)H_D^-(x,z)$ is introduced.

Theorem 5.1 *If u(x,z) is a classical solution to one of the boundary value problems and there exist positive* $\varepsilon_1, \varepsilon_2$ *such that.*

(*for an internal BVP*)

$$|g(x, z)| = O\left(|z|^{N-1-\varepsilon_1}\right) \, at \, |z| \to \infty,$$

(*for an external BVP*)

$$|g(x, z)| = O\left(\|(x, z)\|^{-2-\varepsilon_2}\right) at \|(x, z)\| \to \infty, \quad (x, z) \in D^- + D,$$

then $\hat{u}(x, z)$ *satisfies generalized Green's formula in* $D'(R^{N+1})$:

$$\hat{u}(x, z) = \hat{g} * U + m^2 p_D(x, z)\delta_D * U - u(x, z)n_i(x)\delta_D * U,_i \ . \tag{5.22}$$

Proof Using the rule of differentiation of generalized functions, we easily obtain in $D'(R^{N+1})$

$$\Delta_N^+ \hat{u} + \hat{g} + m^2 p_D \delta_D - (u(x, z)n_i(x)\delta_D),_i = 0.$$

Based on the properties of Green's function and Lemma 3.1, we find

$$\hat{u}(x, z) = \hat{g} * U + m^2 p_D \delta_D * U) - (u(x, z)n_i \delta_D),_i *U.$$

Transferring the differentiation to U in the last term, by virtue of the convolution differentiation property, we obtain (5.22). All convolutions exist due to the boundedness of the carrier u with respect to x and the asymptotic properties of U, u, g at infinity. The theorem is proven.

Its corollary is the following theorem.

Theorem 5.2 *If u(x,z) satisfies the Lipschitz condition on D, i.e.,* $\exists \beta > 0$: $|u(x, z) - u(y, t)| < const \, \|(x - y, z - t)\|^{\beta}$, *then.*

$$u(x, z)H_S^-(x) = \int_{-\infty}^{+\infty} dt \int_{S^-} U(y - x, t - z)g(y, t)\, dV(y) +$$

$$+m^2 \int_{-\infty}^{+\infty} dt \int_S U(y - x, t - z)\frac{du}{dn(y)} dS(y) + \text{V.P.} \int_{-\infty}^{+\infty} dt \int_S u(y)T(y - x, t - z, n(y))\, dS(y).$$

Proof If we go to the integral notation in formula (5.22) and use the symmetry properties, we obtain the formula of the theorem. For $x \notin D$ all the integrals are regular, since the integrands have no singularities in D, and also by virtue of (5.2)–(5.4) and the asymptotic properties of the fundamental solutions on ∞. The formula expresses the solution within the region through its boundary values and the values of the normal derivative, i.e., it is similar to Green's formula for the Laplace equation, and for $m = 1$ (the case that corresponds to the static problem, $v = 0$) it just coincides with it (p.338 [4]).

Let us prove this formula on the boundary, taking into account the definition and using Lemma 5.3.

Let $(x^*, z^*) \in D,\ (x, z) \in D^-,\ (x, z) \to (x^*, z^*)$. Then we have

$$u(x^*, z^*) = \lim_{(x,z)\to(x^*,z^*)} u(x, z) = \lim_{(x,z)\to(x^*,z^*)} \int_{-\infty}^{+\infty} dt \int_{S^-} \left[U(y - x, t - z)g(y, t)\right]dV(y) +$$

$$\lim_{(x,z)\to(x*,z*)} \int_{-\infty}^{+\infty} dt \int_{S} m^2 U(y - x, t - z)\frac{du}{dn(y)} dS(y) +$$

$$+ \lim_{(x,z)\to(x*,z*)} \int_{-\infty}^{+\infty} dt \int_{S} T(y - x, t - z, n(y))u(y)dS(y).$$

The first two integrals are continuous in *(x,z)*, due to the weak singularity of the integrands, and the third integral can be transformed using an analog of the Gauss formula (Lemma 3.1):

$$u(x^*,z^*) = \int\limits_{-\infty}^{+\infty} dt \int\limits_{S^-} \left[U(y-x^*, t-z^* g(y,t)\right]dV(y)+$$

$$+\int\limits_{-\infty}^{+\infty} dt \int\limits_{S} m^2 U(y-x^*, t-z^*)\frac{du}{dn(y)} dS(y)+$$

$$+\lim_{(x,z)\to(x*,z*)} \int\limits_{-\infty}^{+\infty} dt \int\limits_{S} (u(y,t)-u(x^*,z^*))T(y-x, t-z, n(y))dS(y)+$$

$$+u(x^*,z^*) \lim_{(x,z)\to(x*,z*)} \int\limits_{-\infty}^{+\infty} dt \int\limits_{S} T(y-x, t-z, n(y)dS(y) = ...+$$

$$+\int\limits_{-\infty}^{+\infty} dt \int\limits_{S} (u(y,t)-u(x^*,z^*))T(y-x^*, t-z^*, n(y))dS(y) + u(x^*,z^*) = ...+$$

$$+\text{V.P.} \int\limits_{-\infty}^{+\infty} dt \int\limits_{S} u(y,t)T(y-x^*, t-z^*, n(y)dS(y) -$$

$$-u(x^*,z^*)\text{V.P.} \int\limits_{-\infty}^{+\infty} dt \int\limits_{S} T(y-x^*, t-z^*, n(y))dS(y) + u(x^*,z^*) = ...+$$

$$+\text{V.P.} \int\limits_{-\infty}^{+\infty} dt \int\limits_{S} u(y,t)T(y-x^*, t-z^*, n(y))dS(y) + (1-0,5)u(x^*,z^*)$$

$$+\text{V.P.} \int\limits_{-\infty}^{+\infty} dt \int\limits_{S} (u(y,t)-u(x^*,z^*))T(y-x^*, t-z^*, n(y))dS(y) - u(x^*,z^*) = ...+$$

Here we used the continuity of the third integral in the first equality, which allows us to write it on the boundary as an integral with a weak singularity due to the Lipschitz condition for *u(y,t)*, and then the property of the integral in the sense of the principal value, which for integrable functions coincides with the value Lebesgue integral. Moving $0,5u(x^*,z^*)$ to the left side, we obtain the formula of the theorem for boundary points. The theorem is proven.

5.5 Integral Representation of Solutions of Three-Dimensional Transport Boundary Value Problems for Subsonic Speeds

We give an integral representation of the subsonic boundary value problem in three-dimensional space to the case that in Theorem 4.2 corresponds to $N = 2$. Moreover, as follows from formulas (5.13), (5.16), we have

$$\hat{U}(x,z) = \frac{1}{4\pi\sqrt{m^2\|x\|^2 + z^2}},$$
$$\hat{T}(x,z,v) = -\frac{m^2}{4\pi}\frac{v_3 z + m^2(v,x)}{\left(z^2 + m^2\|x\|^2\right)^{3/2}}.$$

Therefore, the solution to the problem has the form:
for $x \in S^-$

$$\begin{aligned}4\pi u(x,z) &= \int_{-\infty}^{+\infty} dt \int_{S^-} \frac{g(y,t)}{\sqrt{m^2\|y-x\|^2 + (t-z)^2}} dy_1 dy_2 + \\ &+ m^2 \int_{-\infty}^{+\infty} dt \int_{S} \frac{1}{\sqrt{m^2\|y-x\|^2 + (t-z)^2}} \frac{du}{dn(y)} dS(y) + \\ &+ m^4 \int_{-\infty}^{+\infty} dt \int_{S} \frac{(n(y), y-x)u(y)}{\left((t-z)^2 + m^2\|y-x\|^2\right)^{3/2}} dS(y);\end{aligned} \tag{5.23}$$

for $x \in S$

$$\begin{aligned}2\pi u(x,z) &= \int_{-\infty}^{+\infty} dt \int_{S^-} \frac{g(y,t)}{\sqrt{m^2\|y-x\|^2 + (t-z)^2}} dy_1 dy_2 + \\ &+ m^2 \int_{-\infty}^{+\infty} dt \int_{S} \frac{1}{\sqrt{m^2\|y-x\|^2 + (t-z)^2}} \frac{du}{dn(y)} dS(y) + \\ &+ \text{V.P.} \int_{-\infty}^{+\infty} dt \int_{S} \frac{m^4(n(y), y-x)u(y)}{\left((t-z)^2 + m^2\|y-x\|^2\right)^{3/2}} dS(y).\end{aligned} \tag{5.24}$$

It is easy to see that for $(x,z) \in D$ the formula is a singular boundary integral equation for solving a boundary value problem with the Neumann boundary condition. For Dirichlet conditions, the same formula is the Fredholm integral equation of the first kind with a polar core. Solving the BIE (5.24) on the boundary, we determine the unknown boundary functions, after which the formula (5.22) gives a solution to the posed boundary value problem.

5.6 Integral Representation of Solutions of Plane Transport Boundary Value Problems for Subsonic Speeds

A plane boundary value problem corresponds to $N = 1$. In this case $S^- = \{x \in R^1 : x \in (a, b)\}$, $S = \{x \in R^1 : x = a \cup x = b\}$, that is, the domain of definition with respect to x is a segment, and the boundary consists of its two boundary points. This does not allow us to use the formula (5.22) of Theorem 4.1 to construct a solution. We represent the corresponding generalized function and its derivatives in the form:

$$
\begin{gathered}
\hat{u}(x,z) = u(x,z)H_D^-(x,z) = u(x,z)H(x-a)H(b-x)1(z), \\
\hat{u}_{,z} = u_{,z}\, H_D^-, \qquad \hat{u}_{,zz} = u_{,zz}\, H_D^-, \\
\hat{u}_{,x} = u_{,x}\, H_D^- + u(a,z)\delta(x-a)H(b-a) - u(b,z)H(b-a)\delta(b-x) = \\
= u_{,x}\, H_D^- + u(a,z)\delta(x-a) - u(b,z)\delta(b-x), \\
\hat{u}_{,xx} = u_{,xx}\, H_D^- + u_{,x}(x,z)\delta(x-a) - u_{,x}(b,z)\delta(b-x) + \\
+u(b,z)\delta'(x-a) - u(b,z)\delta'(b-x) \quad .
\end{gathered}
$$

The equation in the space of generalized functions has the form

$$
\begin{aligned}
&\Delta_N^+\hat{u} + \hat{g}(x,z) = u_{,x}(x,z)\delta(x-a) \\
&-u_{,x}(b,z)\delta(b-x) + u(b,z)\delta'(x-a) - u(b,z)\delta'(b-x)
\end{aligned}
$$

from which follows

$$
\begin{aligned}
\hat{u} = -\hat{g}(x,z) * \hat{U} - \{u_{,x}(a,z)\delta(x-a) - u_{,x}(b,z)\delta(b-x) + \\
+u(a,z)\delta'(x-a) - u(b,z)\delta'(b-x)\} * \hat{U} = \\
= -\hat{g}(x,z) * \hat{U} - u_{,x}(a,z)) \underset{z}{*} \hat{U}(x-a,z) + u_{,x}(b,z) \underset{z}{*} \hat{U}(x-b,z) + \\
-u(a,z) \underset{z}{*} \hat{U}_{,x}(x-a,z) + u(b,z) \underset{z}{*} \hat{U}_{,x}(x-b,z),
\end{aligned}
\tag{5.25}
$$

where the sign $\underset{z}{*}$ means incomplete convolution only in z. Here we used the convolution properties with the delta function and the convolution shift property. The integral representation of this formula has the form:

$$\hat{u} = -\int_{-\infty}^{\infty} d\zeta \int_{a}^{b} g(y,\zeta)\, U(x-y, z-\zeta)dy-$$

$$-\int_{-\infty}^{\infty} \{u_{,x}(a,\zeta)U(x-a,z-\zeta) - u_{,x}(b,\zeta)U(x-b,z-\zeta)\}d\zeta + \tag{5.26}$$

$$-\int_{-\infty}^{\infty} \{u(a,\zeta)U_{,x}(x-a,z-\zeta) - u(b,\zeta)U_{,x}(x-b,z-\zeta)\}d\zeta.$$

Given the type of fundamental solution, we obtain

$$2m\pi u(x,z)H(x-a)H(b-x) = \int_{-\infty}^{\infty} d\zeta \int_{a}^{b} g(y,\zeta)\, \ln\left(\frac{\sqrt{(z-\zeta)^2+m^2(x-y)^2}}{m}\right)dy-$$

$$-\int_{-\infty}^{\infty} u_{,x}(a,\zeta)\ln\left(\frac{\sqrt{(z-\zeta)^2+m^2a^2}}{m}\right)d\zeta + \int_{-\infty}^{\infty} u_{,x}(b,\zeta)\ln\left(\frac{\sqrt{(z-\zeta)^2+m^2b^2}}{m}\right)d\zeta + \tag{5.27}$$

$$+m^2(x-a)\int_{-\infty}^{\infty}\frac{u(a,\zeta)d\zeta}{(z-\zeta)^2+m^2(x-a)^2} - m^2(x-b)\int_{-\infty}^{\infty}\frac{u(b,\zeta)}{(z-\zeta)^2+m^2(x-b)^2}d\zeta.$$

For internal points, all integrands are continuous; they decay at infinity due to the form of kernels and the decay conditions at infinity of the solution to the problem. The formula describes the solution of the problem for known values of the function and its derivatives at the ends of the segment. However, at the ends of the segment, the last integral is singular; integrand functions have singularities at points. We study the limiting properties of this integral near these boundary points. To do this, we consider the last integrals of the solution at near points. In particular, at the point $x = (a+\varepsilon)$:

$$\varepsilon\int_{-\infty}^{\infty}\frac{u(a,\zeta)d\zeta}{(z-\zeta)^2+m^2\varepsilon^2} = \varepsilon\int_{|z-\zeta|>1}\frac{u(a,\zeta)d\zeta}{(z-\zeta)^2+m^2\varepsilon^2}+$$

$$+\varepsilon\int_{|z-\zeta|\le 1}\frac{u(a,\zeta)d\zeta}{(z-\zeta)^2+m^2\varepsilon^2} = \varepsilon\int_{|y|>1}\frac{u(a,z-y)dy}{y^2+m^2\varepsilon^2} + \varepsilon\int_{|y|<1}\frac{u(a,z-y)dy}{(z-y)^2+m^2\varepsilon^2}. \tag{5.28}$$

When passing to the limit with respect to $\varepsilon \to +0$, the first term on the right-hand side tends to zero, since the integral exists, since the integrands are continuous for $|y| > 1$ and tend to zero at infinity.

Let us consider the second term. Passing to the limit in it, we obtain

$$\varepsilon \lim_{\varepsilon\to+0} \int\limits_{|y|<1} \frac{u(a,z-y)dy}{y^2+m^2\varepsilon^2} = \varepsilon \lim_{\varepsilon\to+0} \int\limits_{\varepsilon<|y|<1} \frac{(u(a,z-y))dy}{y^2+m^2\varepsilon^2} =$$

$$= \lim_{\varepsilon\to+0} \int\limits_{\varepsilon<|y|<1} \frac{(u(a,z-y)-u(a,z)-u_{,z}(a,z)y)dy}{y^2+m^2\varepsilon^2} +$$

$$+u(a,z) \lim_{\varepsilon\to+0} \int\limits_{\varepsilon<|y|<1} \frac{\varepsilon dy}{y^2+m^2\varepsilon^2} + \varepsilon u_{,z}(a,z) \lim_{\varepsilon\to+0} \int\limits_{\varepsilon<|y|<1} \frac{ydy}{y^2+m^2\varepsilon^2} =$$

$$= \varepsilon \text{V.P.} \int\limits_{|y|<1} \frac{(u(a,z-y)-u(a,z)-u_{,z}(a,z)y)dy}{y^2} + m u(a,z) \lim_{\varepsilon\to+0} \operatorname{arctg}\frac{y}{m\varepsilon}\Big|_{-\frac{1}{m\varepsilon}}^{\frac{1}{m\varepsilon}} =$$

$$= m\pi\, u(a,z).$$

Transferring $m\pi\, u(a,z)$ to the right side, we obtain the same formula, taking into account the definition of the characteristic function on the boundary ($H(0) = 1/2$). Similarly, the formula is proven for $x = b$.We formulate the result as a theorem.

Theorem 5.3 *The solution of a plane transport boundary value problem at subsonic speeds satisfies the boundary integral equations of the form:*

$$m\pi u(a,z) = \int\limits_{-\infty}^{\infty} d\zeta \int\limits_{a}^{b} g(y,\zeta) \ln\left(\frac{\sqrt{(z-\zeta)^2+m^2(a-y)^2}}{m}\right) dy -$$

$$- \int\limits_{-\infty}^{\infty} u_{,x}(a,\zeta) \ln\left(\frac{\sqrt{(z-\zeta)^2+m^2a^2}}{m}\right) d\zeta + \int\limits_{-\infty}^{\infty} u_{,x}(b,\zeta) \ln\left(\frac{\sqrt{(z-\zeta)^2+m^2b^2}}{m}\right) d\zeta +$$

$$+m^2(b-a) \int\limits_{-\infty}^{\infty} \frac{u(b,\zeta)}{(z-\zeta)^2+m^2(b-a)^2} d\zeta$$

$$m\pi u(b,z) = \int\limits_{-\infty}^{\infty} d\zeta \int\limits_{a}^{b} g(y,\zeta) \ln\left(\frac{\sqrt{(z-\zeta)^2+m^2(b-y)^2}}{m}\right) dy -$$

$$- \int\limits_{-\infty}^{\infty} u_{,x}(a,\zeta) \ln\left(\frac{\sqrt{(z-\zeta)^2+m^2a^2}}{m}\right) d\zeta + \int\limits_{-\infty}^{\infty} u_{,x}(b,\zeta) \ln\left(\frac{\sqrt{(z-\zeta)^2+m^2b^2}}{m}\right) d\zeta +$$

$$+m^2(b-a) \int\limits_{-\infty}^{\infty} \frac{u(a,\zeta)d\zeta}{(z-\zeta)^2+m^2(b-a)^2}.$$

This system of two equations allows us to determine two unknown boundary functions, after which, using formulas (5.27), we determine the solution of the boundary value problem.

Note that the kernels of these integrals are continuous functions. For the first boundary value problem, this is a system of equations of Fredholm type of the second kind, and for the second boundary value problem it is a Fredholm type of the first kind. Since the integrals are improper and have a logarithmic kernel, for their existence it is enough to require that for the Neumann conditions the relations

$$u_{,x}(a,z)\ln|z| = O\left(|z|^{-(2+\alpha)}\right), \quad \forall\alpha > 0,$$
$$u_{,x}(b,z)\ln|z| = O\left(|z|^{-(2+\beta)}\right), \quad \forall\beta > 0, \tag{5.29}$$

and for the Dirichlet conditions, the decay condition of the solution at infinity is sufficient.

5.7 Formulation of Transport Boundary Value Problem for Supersonic Speeds: Uniqueness of the Solution

For $M > 1$, Eq. (5.6) is a classical wave equation of strictly hyperbolic type:

$$\Delta_N u - m^2 u_{,zz} + g(x,z) = 0, m^2 = M^2 - 1. \tag{5.30}$$

Consequently, the solution may have discontinuities of derivatives on characteristic surfaces on which $m^2 \nu_z^2 - \nu_i \nu_i = 0$. Fracture surface F, $\nu = (\nu_1, \ldots, \nu_N, \nu_z)$ is the unit normal vector to F in R^{N+1}, $\nu_z = \nu_{N+1}$. At the fronts of solutions of the wave equation, the Hadamard conditions for jumps are

$$[u]_F = 0, \quad \left[\nu_z u_{,j} - \nu_j u_{,z}\right]_F = 0, \tag{5.31}$$

$$\left[\nu_j u_{,j} - M^2 \nu_z u_{,z}\right]_F = 0, \quad j = 1, ..., N. \tag{5.32}$$

The second condition, the continuity of the tangent derivatives at the fronts in (5.31), is a consequence of the continuity of the solution. Condition (5.32) is the law of conservation of momentum at wave fronts.

Definition We call a wave *a weak shock*, if $\left[\nu_j u_{,j}\right]_F = 0$, $j = 1, ..., N + 1$, is a *strong shock*, if $\left[\nu_j u_{,j}\right]_F = \infty$. Otherwise, we simply have a *shock* wave.

The solution of Eqs. (5.30) with a finite number of wave fronts that are sufficiently smooth almost everywhere and dividing the domain of definition into a finite or countable number of subdomains, where the solution is twice differentiable in all variables, will be called *classical.*

In physical problems, traffic loads are usually finite. At supersonic speeds, excitation in the medium propagates more slowly than them, so fronts arise that separate the unperturbed zone, where $u = 0$, and the perturbed one. We denote $C = \{(x,z) \in D : z \geq 0\}$, $C^- = \left\{(x,z) \in D^- : z \geq 0\right\}$:

$$\text{supp}\ u_D(x,z) \subset C, \quad \text{supp}\ u(x,z) \subset C + C^-. \tag{5.33}$$

We denote $u(x, 0) = u^0(x)$, $u_z(x, 0) = u^0{,_z}(x)$. If the leading edge is a weak shock wave, then

$$u^0(x) = 0, \quad u^0{,_z}(x) = 0. \tag{5.34}$$

Theorem 5.4 *If $u(x, z)$ is a classical solution to the first (second) internal boundary value problem, then it is unique.*

Proof By virtue of the linearity of the problem, it suffices to prove the uniqueness of the zero solution to the homogeneous boundary value problem with the same asymptotic behavior at infinity. Let $u(x,z)$ be a solution of the equation for $g \equiv 0$ satisfying zero boundary conditions. Multiplying (5.6) by $u,_z$, after simple calculations we get.

$$(u_j, u_z),_i - 0.5(u_i u_i - M^2 u,_z u,_z),_z = 0, j = 1, \ldots, N.$$

We integrate this relation over a bounded cylinder $C^-_{ab} = D^- \cap \{(x, z) : 0 \le a < z < b\}$, C_{ab} being its boundary on *D*. Given the conditions of the lemma, we also obtain

$$\int_{S^-} dS(x) \int_a^b ((u_i, u_z),_i - 0.5(u_i u_i - M^2 u,_z u,_z))dz = 0.$$

The integral cannot be immediately converted using the Ostrogradsky–Gauss formulas, since the integrands are discontinuous at the wave fronts. Let $F_k(x')$ be such fronts, h^k being respectively normal to them. Taking into account the separation of the integration domain by wave fronts, using the above theorem for the subdomains between them, we reduce the last equality to the

$$\int_{C_{ab}} \sum_{i=1}^{N} u,_z u,_i n_i dS(x, z) + \int_F \sum_{i=1}^{N+1} \left[u,_z u,_i h_i^k - 0,5(u,_i u,_i + M^2 u,_z^2) h_z^k\right]_{F_k} dS(x) +$$
$$+ \int_F \left[u,_z^2 - 0,5(u,_i u,_i + M^2 u,_z u,_z)\right]_{(x,a)}^{(x,b)} dS(x) = 0. \tag{5.35}$$

The first integral is equal to zero due to zero boundary conditions. The jumps on the fronts in the second integral can be transformed taking into account the conditions on the fronts, where the expressions in square brackets are continuous:

$$\sum_{i=1}^{N+1}\left[u,_z u,_i h_i^k - 0,5(u,_i u,_i + M^2 u,_z^2)h_z^k\right]_{F_k} =$$
$$= 0,5\left[(u,_z h_i^k - u,_i h_z^k)u,_i + (u,_i h_i^k - M^2 u,_z h_z^k)u,_z\right]_{F_k} =$$
$$= 0,5\left[(u,_z^- h_i^k - u,_i^- h_z^k)[u,_i]_{F_k} + (u,_i^- h_i^k - M^2 u,_z^- h_z^k)[u,_z]_{F_k}\right] =$$
$$= 0,5\left[\left[(u,_z h_i^k - u,_i h_z^k)\right]_{F_k} u,_i^- + \left[u,_i h_i^k - M^2 u,_z h_z^k\right]_{F_k} u,_z^-\right] = 0.$$

Here, the symbol "-" marks the values of the corresponding expression behind the wave front. The last integral is converted to

$$\int_{S^-}\left[\sum_{i=1}^{N} u,_i u,_i - (M^2-1)u,_z^2\right]_{(x,a)} dS(x) = \int_{S^-}\left[\sum_{i=1}^{N} u,_i u,_i - (M^2-1)u,_z^2\right]_{(x,b)} dS(x).$$

Since $M > 1$, the expression under the sign of each integral is positive. Further reasoning is similar to Lemma 2.1. The theorem is proven.

The proof of the theorem for the external problem is similar, you only need to introduce a bounding cylinder of a sufficiently large radius and take into account the asymptotic behavior of the solutions at infinity.

5.8 Dynamic Analogue of Green's Formula in $D'(R^{N+1})$ at Supersonic Speeds

We introduce regular generalized functions defined on R^{N+1} with the same carrier:

$$\hat{u}(x,z) = u(x,z)H_S^-(x)H(z),\ \hat{g}(x,z) = g(x,z)H_S^-(x)H(z).$$

Let $U_0(x,z)$ be the Green function of Eq. (5.30) satisfying the conditions:

$$\Delta_N U_0 - m^2 U_0 + \delta(x,z) = 0, \tag{5.36}$$

$$\text{supp}\, U_0(x,z) \subset K^+(x,z),\ U_0(x,z) \to 0 \text{ at } \|(x,z)\| \to \infty. \tag{5.37}$$

Here $K^+(x,z) = \{(x,z) : z > m\|x\|\}$ is the cone of perturbations generated by the source at the point (0,0), $K(x,z)$ is its lateral surface, and $\mathbf{K} = K^+ + K$. Further, the differential operator in (5.30) is denoted by ν_m^-. Along with U_0, we introduce functions antiderivative in z

$$\begin{aligned} U_k(x,z) &= U_{k-1}(x,z) * H(z)\delta(x) = U_{k-1}(x,z) \underset{z}{*} H(z),\\ T_k(x,z,n) &= U_k,_i(x,z)n_i = T_{k-1} \underset{z}{*} H(z), \qquad k = 1,2\ldots \end{aligned} \tag{5.38}$$

It is easy to see that, due to the semi-boundedness of the supports U and $H(z)$, all convolutions exist, have the same carrier $\boldsymbol{K}$, and satisfy the equations:

$$\begin{aligned} \nu_m^- U_{k+1} + \frac{1}{k!} z^k H(z)\delta(x) = 0, \\ \nu_m^- T_{k+1} + \frac{1}{k!}\frac{d}{dn}\left[z^k H(z)\delta(x)\right] = 0. \end{aligned} \tag{5.39}$$

Theorem 5.5 *A generalized function* $\hat{u}(x, z)$, *where* $u(x, z)$ *is the classical solution of a boundary value problem, can be represented as the sum of following convolutions:*

$$\begin{aligned} &\hat{u} = p_D(x, z)\delta_S(x)H(z) * U_0(x, z) + u_D,_z(x, z)n_i(x)\delta_S(x)H(z) * U_1,_i(x, z) + \\ &+ u^0(x, z)n_i(x)\delta_S(x) \underset{x}{*} U_1,_i(x, z) \\ &+ (H_S^-(x)u^0(x) \underset{x}{*} U_0(x, z)),_z + m^2 u^0,_z(x)H_S^-(x) \underset{x}{*} U_0(x). \end{aligned} \tag{5.40}$$

Proof Differentiating $\hat{u}$ according to the rules for generalized functions, taking into account relations at the fronts, we have

$$\begin{aligned} \nu_m^- \hat{u}(x, z) + \hat{g}(x, z) = p_D(x, z)\delta_S(x)H(z) + [n_i(x)\delta_S(x)H(z)u_D(x, z)],_i + \\ + m^2 u^0,_z(x)H_S^-(x)\delta_S(z) + m^2\left[H_S^-(x)\delta_S(z)u^0(x)\right],_z. \end{aligned}$$

Using the property of Green's function, we find

$$\begin{aligned} \hat{u}(x, z) = U_0(x, z) * p_D(x, z)\delta_S(x)H(z) + [n_i(x)\delta_S(x)H(z)u_D(x, z)],_i * U_0 + \\ + m^2 u^0,_z(x)H_S^-(x) \underset{z}{*} U_0(x, z) + m^2\left[H_S^-(x)u^0(x) \underset{z}{*} U_0(x, z)\right],_z. \end{aligned} \tag{5.41}$$

Here $a(x)\delta_s(x)H(z)$ is a simple layer on the "semi-infinite cylinder" C, and $\beta(x, z)H_S^-(x)\delta(z)$ is a simple layer on its base $\left\{S^- \times 0\right\}$:

$$\beta(x, z)H_S^-(x)\delta(z) * f(x, z) = \beta(x, z)H_S^-(x)\delta(z) \underset{z}{*} f(x, z).$$

Since $U_0,_z = U_1$, it is easy to see that

$$\begin{aligned} [u(x, z)n_i(x)\delta_S(x)H(z)],_i * U_0 = [u(x, z)n_i(x)\delta_S(x)H(z) * U_{1'z}],_i = \\ = u,_z(x, z)n_i(x)\delta_S(x)H(z) * U_{1'i} + u^0(x)n_i(x)\delta_S(x) \underset{z}{*} U_{1'i}. \end{aligned}$$

Substituting this into (5.41), we obtain the formula of the theorem. Moreover, due to the convolution property, the solution is unique. The theorem is proven.

In contrast to Green's formula, the resulting formula expresses the solution inside the domain in terms of the boundary values of the normal and tangent derivatives of U by D, as well as the values of $u(x,0)$ и $u_{,z}(x,0)$, which is quite natural for the hyperbolic problem; here these last conditions are similar to the initial ones for solving the wave equations in $R^{N+1} \times t$.

Note that the formula of the theorem is preferable to (5.41), since it expresses the function $u(x,z)$ in terms of more regular derivatives of the antiderivative Green's function $\partial_j U_1$, but not $\partial_j U_0$. Further, we show for different N its convenience in constructing integral representations of the solution of the boundary value problem.

5.9 Dynamic Analogue of Gauss Formula at Supersonic Speeds

In contrast to the subsonic case, where the representation of the characteristic function of the domain for determining the solution of the equation using the fundamental solution of the equation has the form similar to the Gauss formula for the fundamental solution of the Laplace equation, in the supersonic case the formal recording of this formula in integral form for spaces of different dimensions leads to divergent integrals. However, the introduction of the antiderivative Green's function allows one to obtain an analog of this formula, which is here called *dynamic*.

Lemma 5.4 *Functions $U_k(x,z)$ satisfy the equalities:*

$$U_{k+2'i}(x,z) \underset{x}{*} n_i(x)\delta_S(x) + m^2 U_k \underset{x}{*} H_S^-(x) - \frac{1}{(k+1)!} z^{k+1} H(z) H_S^-(x) = 0,$$
$$U_{k+1'i}(x,z) \underset{x}{*} n_i(x)\delta_S(x) + m^2 U_{k,z} \underset{x}{*} H_S^-(x) + \frac{1}{k!} z^k H(z) H_S^-(x) = 0. \tag{5.42}$$

Proof We collapse the Eq. (5.39) with $H_S^-(x)H(z)$ and use the property of convolution differentiation:

$$\nu_m^- U_{k+1} * H_S^-(x)H(z) = \Delta_N U_{k+1} * H_S^-(x)H(z) - m^2 U_{k+1,zz} * H_S^-(x)H(z) =$$
$$= U_{k+2,i} \underset{x}{*} n_i(x)\delta_S(x) - m^2 U_{k+1,z} \underset{x}{*} H_S^-(x) =$$
$$= -\frac{1}{k!} z^k H(z)\delta(x) * H_S^-(x)H(z) = \frac{1}{(k+1)!} z^{k+1} H(z) * H_S^-(x).$$

Since the carrier of functions with respect to z is bounded on the left, all convolutions exist. This implies the first formula of the lemma. Convolution (5.39) with $\delta(z)H_S^-(x)$ by z gives the second formula. The lemma is proven.

5.10 Integral Representations of Solutions of Three-Dimensional Boundary Value Problems for Supersonic Speeds

In this case N = 2 and the fundamental solution of wave equation having the form:

$$U_o = \frac{1}{2\pi}\frac{H(z - m\|x\|)}{\sqrt{z^2 - m^2\|x\|^2}}, x = (x_1, x_2) \tag{5.43}$$

represents a strong shock wave with a front $z = m\|x\|$, the carrier of which is the inside of the cone: $m\|x\| < z$. Calculating by formula (5.43), we find the necessary antiderivatives and their derivatives in the direction with the same carrier:

$$U_1 = \frac{H(z - m\|x\|)}{2\pi}\ln\frac{z + \sqrt{z^2 - m^2\|x\|^2}}{m\|x\|}, \tag{5.44}$$

$$T_1 = \frac{1}{2\pi}\frac{H(z - m\|x\|)}{\sqrt{z^2 - m^2\|x\|^2}}\left[n_z - \frac{z}{\|x\|}\frac{d\|x\|}{dn}\right], \tag{5.45}$$

$$T_2 = \frac{H(z - m\|x\|)}{2\pi}\left[n_z \ln\frac{|z| + \sqrt{z^2 - m^2\|x\|^2}}{m\|x\|} - \frac{1}{\|x\|}\frac{d\|x\|}{dn}\sqrt{z^2 - m^2\|x\|^2}\right]. \tag{5.46}$$

We write the formulas of Lemma 5.4 and Theorem 5.5 taking these representations into account.

Lemma 5.5 *The dynamic analogue of the Gauss formula at supersonic speeds has the form:*

$$\text{V.P.}\int\limits_{S_x(z)} \frac{\sqrt{z^2 - m^2r^2}}{r}\frac{dr}{dn(y)}dS(y) + m^2\int\limits_{S_x(z)} \frac{dV(y)}{\sqrt{z^2 - m^2r^2}} + 2\pi\, z_+ H_S^-(x) = 0,$$

$$V.P.\int\limits_{S_x(z)} \frac{z}{r}\frac{dr}{dn(y)}\frac{dS(y)}{\sqrt{z^2 - m^2r^2}} + m^2\frac{d}{dz}\int\limits_{S_x(z)} \frac{dV(y)}{\sqrt{z^2 - m^2r^2}} + 2\pi H(z)H_S^-(x) = 0,$$

where $S_x(z) = \{y \in S : \; mr < z\}$, $S_x^-(z) = \left\{y \in S^- : \; mr < z\right\}$, $r = \|x - y\|$.

Proof The integral representation of the formulas of Lemma 9.1 has the form:

$$\int\limits_S T_{k+2}(y, x, z, n(y))dS(y) + m^2\int\limits_{S^-} U_k(x, y, z)dV(y) = \frac{1}{(k+1)!}z^{k+1}H(z)H_S^-(x), \tag{5.47}$$

$$\int_S T_{k+1}(y,x,z,n(y))dS(y) + m^2 \int_{S^-} \partial_z U_k(x,y,z)dV(y) = \frac{1}{k!} z^k H(z) H_S^-(x). \tag{5.48}$$

It is easy to see that for $x \in S$, all integrals exist due to weak singularities of the integrands at the front $z = mr$. It is clear that if $z < 0, \quad x \in S$, then $S_x^-(z)$ is an empty set; therefore, the left and right sides of the equality are 0.

For $k = 0$, taking into account the normal property on D: $n_z = 0$, we obtain the formulas of the lemma. Note that the first formula of the lemma can be obtained from the second by simple integration over z from 0 to z.

Let us show that the equalities are also preserved for $x \in S$, if we take into account the definition of $H_S^-(x)H(z)$. We denote

$$O_\varepsilon(x) = \{y \in S: \ \|x-y\| < \varepsilon\}, \quad S_\varepsilon = S - O_\varepsilon, \quad Ш_\varepsilon(x) = \{y: \ \|x-y\| < \varepsilon\},$$

$Ш_\varepsilon^\pm(x) = Ш_\varepsilon \cap S^\pm$, $S_\varepsilon^\pm = S^- \pm Ш_\varepsilon^\pm$, $\Gamma_\varepsilon^\pm$ are boundaries of $Ш_\varepsilon^\pm$ in $S^\pm$ accordingly " + " and "-". Lemma 5.4 implies

$$\int_{S_\varepsilon + \Gamma_\varepsilon^+} \frac{zH(z-mr)}{r} \frac{dr}{dn(y)} \frac{dS(y)}{\sqrt{z^2 - m^2 r^2}} + m^2 \frac{d}{dz} \int_{S_\varepsilon^+} \frac{H(z-mr)dV(y)}{\sqrt{z^2 - m^2 r^2}} + 2\pi = 0,$$

$$\int_{S_\varepsilon + \Gamma_\varepsilon^-} \frac{zH(z-mr)}{r} \frac{dr}{dn(y)} \frac{dS(y)}{\sqrt{z^2 - m^2 r^2}} + m^2 \frac{d}{dz} \int_{S_\varepsilon^-} \frac{H(z-mr)dV(y)}{\sqrt{z^2 - m^2 r^2}} = 0.$$

Due to the fact that on $\Gamma_\varepsilon^\pm$: $\frac{dr}{dn(y)} = \pm 1$ according to the sign, adding up these two equalities and dividing by 2, we obtain

$$\int_{S_\varepsilon(x)} \frac{z}{r} \frac{dr}{dn(y)} \frac{H(z-mr)}{\sqrt{z^2-m^2r^2}} dS(y) + m^2 \frac{d}{dz} \int_{S_\varepsilon^-} \frac{H(z-mr)dV(y)}{\sqrt{z^2-m^2r^2}} +$$
$$+ m^2 \frac{d}{dz} \int_{Ш_\varepsilon} \frac{H(z-mr)dV(y)}{\sqrt{z^2-m^2r^2}} + \pi = 0. \tag{5.49}$$

The last integral for small ε ($\varepsilon < z/m$) is easily calculated if we go to the polar coordinate system:

$$I = \int_0^{2\pi} d\theta \int_0^\varepsilon \frac{r dr d\theta}{\sqrt{z^2 - m^2 r^2}} = \frac{m\varepsilon^2}{z}, \ \frac{dI}{dz} = -\frac{m\varepsilon^2}{z^2} \rightarrow_{\varepsilon \to 0} 0, z \neq 0.$$

Passing to the limit in $\varepsilon \to 0$, we obtain the second formula of the lemma for $x \in S$. In this case, the first integral is taken in the sense of the principal value, which exists due to the antisymmetry of the integrand. From it, as was said, the first formula follows for the boundary points. The lemma is proven.

If $z > md$, where $d = \max_{y_i \in S} \|y_1 - y_2\|$ is a cylinder diameter, then $S_x(z) = S$, $S_x^-(z) = S^-$. That is, the integration domain becomes constant (independent of z).

Theorem 5.6 (*Dynamic analogue of Green's formula*). *When $M > 1$ the solution of the transport boundary value problem is representable in the form.*

$$2\pi u(x, z) =$$

$$= \int_0^z \left\{ \text{V.P.} \int_{S_{x(z-\zeta)}} \left[u_{,z}(y, \zeta) \frac{z-\zeta}{2} \frac{dr}{dn(y)} + p_D(y, \zeta) \right] \frac{dS(y)}{\sqrt{|z-\zeta|^2 - m^2 r^2}} \right\} d\zeta +$$

$$+ \int_{S_{x(z)}}^{z} \frac{1}{r} \frac{dr}{dn(y)} \frac{u^o(y) dS(y)}{\sqrt{z^2 - m^2 r^2}} + m^2 \int_{S^-_{x(z)}} \frac{\hat{u}^o_{,z} dV(y)}{\sqrt{z^2 - m^2 r^2}} + \frac{d}{dz} \int_{S^-_{x(z)}}^{z} \frac{\hat{u}^o(y) dV(y)}{\sqrt{z^2 - m^2 r^2}}$$

Or, When Changing the Order of Integration,

$$2\pi \hat{u}(x, z)$$

$$= V.P. \int_{S_{x(z)}} dS(y) \left\{ \int_{(mr,z)} \left[u_{,z}(y, \zeta) \frac{z-\zeta}{2} \frac{dr}{dn(y)} + p_D(y, \zeta) \right] \frac{d\zeta}{\sqrt{|z-\zeta|^2 - m^2 r^2}} \right\}$$

$$+ \int_{S_{x(z)}}^{z} \frac{1}{r} \frac{dr}{dn(y)} \frac{u^o(y) dS(y)}{\sqrt{z^2 - m^2 r^2}} + m^2 \int_{S^-_{x(z)}} \frac{\hat{u}_{,z}^o dV(y)}{\sqrt{z^2 - m^2 r^2}} + \frac{d}{dz} \int_{S^-_{x(z)}}^{z} \frac{\hat{u}^o(y) dV(y)}{\sqrt{z^2 - m^2 r^2}}$$

where for $x \in S^-$ the integrals are regular, and for $x \in S$ they are singular, are taken in the sense of the main value.

Proof For $x \notin S$ the formula is a direct consequence of Theorem 8.1, if we write down all convolutions taking into account (5.43)–(5.45). All integrals exist due to the weak convergence of integrands at the fronts. For such x, we can change the order of integration in the first integral:

$$\int_0^z \ldots d\zeta \int_{S_{x(z-\zeta)}} \ldots dS(y) = \int_{S_{x(z)}} \ldots dS(y) \int_{mr}^{z} \ldots dz.$$

If $x \in S$, integrals containing $\frac{1}{r} \frac{dr}{dn(y)}$ are singular. Therefore, to prove the theorem for $x \in S$, here we can use the same formula, but for a region with a punctured neighborhood of x.

So let $x^* \in S$. We transform S in a neighborhood of the point x, bypassing it in a ε-semicircle inside the region S^-, and then we pass to the limit in $\varepsilon \to 0$; as a result, we obtain

$$0 = \int\limits_{S_\varepsilon + \Gamma_\varepsilon^-} \frac{1}{r} \frac{dr}{dn(y)} dS(y) \int\limits_{mr}^{z} \frac{u_{,z}(y,\zeta)(z-\zeta)}{\sqrt{|z-\zeta|^2 - m^2r^2}} d\zeta + \int\limits_{S_\varepsilon + \Gamma_\varepsilon^-} \frac{z}{r} \frac{dr}{dn(y)} \frac{u^o(y)dS(y)}{\sqrt{z^2 - m^2r^2}} + \ldots \tag{5.50}$$

Here, only singular integrals are written out. The rest are continuous in ε at $\varepsilon \to 0$. On $\Gamma_\varepsilon^- : r = \varepsilon$, $\frac{dr}{dn(y)} = -\frac{dr}{dr} = -1$, $dS(y) = \varepsilon d\theta$, where θ is a polar angle with a vertex at a point (x^*, z).

Denote by θ_1, θ_2 the polar angles of the ends of Γ_ε^-, renumbered in the order of the positive direction of the circuit $S_\varepsilon^- + \Gamma_\varepsilon^-$ bypassed in the section $\{S \times z\}$. Notice, that

$$\lim_{\varepsilon\to 0} \int\limits_{S_\varepsilon + \Gamma_\varepsilon^-} \ldots = \lim_{\varepsilon\to 0} \int\limits_{S_\varepsilon} \ldots + \lim_{\varepsilon\to 0} \int\limits_{\Gamma_\varepsilon^-} \ldots$$

We consider the integrals over Γ_ε^-. Given the values of the integrands on Γ_ε^-, they are transformed to

$$I_\varepsilon(x,z) = -\int\limits_{\theta_1}^{\theta_2} d\theta \int\limits_{m\varepsilon}^{z} u_{,z}[x + \varepsilon e(\theta), \zeta](z-\zeta) \frac{d\zeta}{\sqrt{|z-\zeta|^2 - m^2r^2}},$$

$$J_\varepsilon(x,z) = -\int\limits_{\theta_1}^{\theta_2} \frac{zu(x+\varepsilon e(\theta))}{\sqrt{z^2 - m^2\varepsilon^2}} d\theta, \qquad e(\theta) = \frac{y-x}{\|y-x\|}.$$

Since $\lim\limits_{\varepsilon\to 0}(\theta_2 - \theta_1) = -\pi$, therefore,

$$\lim_{\varepsilon\to 0} I_\varepsilon(x,z) = \pi \int\limits_0^z u_{,z}(x,\zeta)d\zeta = \pi\left[u(x,z) - u^o(x)\right],$$

$$\lim_{\varepsilon\to 0} J_\varepsilon(x,z) = \pi u^o(x). \tag{5.51}$$

The integral over S_ε, by definition, coincides with the integral in the sense of the principal value, which exists due to the antisymmetry of the function $\frac{dr}{dn} = \frac{n_i(x_i - y_i)}{r}$ relative to the point x^* in y. Passing to the limit in $\varepsilon \to 0$, in (5.50) with allowance for (5.51), we obtain the formula of the theorem for the boundary points as well. The theorem is proven.

In the case of zero conditions at $z = 0$, the representation of the solution is greatly simplified.

For $x \in S$ the formula gives a boundary integral equation for solving the first or second boundary value problem. Moreover, in the case of the first boundary value problem, in contrast to the subsonic case, on D we need not the function u itself, but its tangent derivative $u,_z$. For the second boundary value problem, the equations are singular.

Unlike the subsonic case, the constructed GIEs are of a non-classical type, since the domain of integration over x depends significantly on z. The kernels of these equations have integrable singularities at the fronts of fundamental solutions and strong singularities at $r = 0$. Studying these types of BIEs is an independent mathematical problem.

5.11 Integral Equations of Plane Supersonic Transport Boundary Value Problem

In this case N $= 1$, the point of space has coordinates (x_1, z). Let the domain of definition strip $D = \{(x, z) : \; |x| < a, -\infty < z < \infty\}, \quad n(x, z) = (\pm 1, 0)$ at $x = \pm a$ according to the sign. We denote $a_1 = a, \quad a_2 = -a, \quad u_i(z) = u(a_i, z)$.

The fundamental solution $U_0(x, z)$ to Eq. (5.30) is the Riemann function:

$$U_0(x, z) = \frac{1}{2m} H(z - m|x|), \qquad U_1 = \frac{1}{2m}(z - m|x|)_+$$

$$T_1 = \frac{1}{2m}(n_2 - mn_1 \text{sgn} x) H(z - m|x), |x = x_1. \tag{5.52}$$

Theorem 5.7 *When $M > 1$ the solution of the transport regional problem in the strip $|x| < a, \; \forall z$, under conditions $u(x, 0) = 0, \;\; u,_z(x, 0) = 0$ has the form.*

$$\begin{gathered} 2m\hat{u} = \sum_{i=1}^{2} (-1)^i H(z - m|x - a_i|) \int\limits_0^{z - m|x - a_i|} u_{i'y}(z) dz- \\ -m \text{ sgn}\,(x - a)\; H(z - m|x - a|)\; u_1(z - m|x - a|)+ \\ +m \text{ sgn}\,(x + a)\; u_2(z - m|x + a|)\; H(z - m|x + a|) \end{gathered} \tag{5.53}$$

and on the boundary ($|x| = a, \;\; \forall z$) the solution satisfies the integral-functional equations with delay

$$
\begin{aligned}
mu_1(z)H(z) &= -H(z)\int\limits_0^z u_{1'y}(\zeta)d\zeta+ \\
&+H(z-2ma)\int\limits_0^{z-2ma} u_{2'y}(\zeta)d\zeta - mu_1(z-2ma)H(z-2ma), \\
mu_2(z)H(z) &= -H(z-2ma)\int\limits_0^{z-2ma} u_{1'y}(\zeta)d\zeta+ \\
&+H(z)\int\limits_0^{z} u_{2'y}(\zeta)d\zeta + mu_1(z-2ma)H(z-2ma).
\end{aligned}
\tag{5.54}
$$

Proof In this case, relations (5.40), taking into account the formulas:

$$H_D^-(x,z) = H(x-a)H(a-x)1(z), \quad \delta_S(x)H(z) = (\delta(x-a)-\delta(a-x))1(z)$$

have the following integral representation:

$$
\begin{aligned}
&2m\hat{u}(x,z) = \\
&= \sum_{i=1}^{2}(-i)^i H(z-m|x-a_i|)\int\limits_0^{z-m|x-a_i|}\left[u_{i,y}(a_i,\zeta) - m\,\text{sgn}\,(x-a_i)u_{i,\zeta}(a_i,\zeta)\right]d\zeta
\end{aligned}
$$

where the zero conditions are taken into account at $z = 0$. Integrating the second term, on the right over ζ we obtain the formulas of the theorem. By virtue of the Le Bois-Reymond lemma, the formulas are valid for $x \neq a_i$. Taking into account the additional definition of the Heaviside function at zero, they are also valid for $x = a_i$ and can be reduced to the form (5.54). We will show it.

To do this, we write equality (5.53) for the point $x^* = (a-\varepsilon, z)$, $0 < \varepsilon < mz$. Since, $x^* \in D^-$, therefore,

$$
\begin{aligned}
&2mu(x-\varepsilon,z) = \\
&= -H(z-m\varepsilon)\int\limits_0^{z-m\varepsilon} u_{,y}(a-\varepsilon,\zeta)d\zeta - mu(a-\varepsilon,z-m\varepsilon)H(z-m\varepsilon)\,\text{sgn}\,(-\varepsilon)+ \\
&+m\,\text{sgn}\,(2a-\varepsilon)\,u_2(z-|2ma-\varepsilon|)H(z-|2ma-\varepsilon|)+ \\
&+H(z-|2ma-\varepsilon|)\int\limits_0^{z-|2ma-\varepsilon|} u_{2,y}(a-\varepsilon,\zeta)d\zeta.
\end{aligned}
$$

Passing in this equality to the limit in $\varepsilon \to 0$, we obtain

$$2mu(a,z) = -H(z)\int_0^z u_y(a,\zeta)d\zeta + mu(a,z)\theta(z) +$$

$$+m\, u_2(z-|2ma|)H(z-|2ma|) + H(z-|2ma|)\int_0^{z-|2ma|} u_{2'y}(a-\varepsilon,\zeta)d\zeta.$$

Whence comes the first formula (5.54). The second Eq. (5.54) is proven similarly. The theorem is proven.

For given $\frac{du}{dx}$ on direct $x = \pm a$, these equations are functional with a "delay in z," since they associate the subsequent boundary values in z with the previous ones. By step-by-step integration over z from 0, they allow us to determine u_1 and u_2 at each subsequent step and, further, by the formulas of the theorem inside the strip.

This completes the construction of the BIE for supersonic speeds, although the formula of Theorem 8.1 allows us to obtain integral representations for $\hat{u}$ for different N, we only need the form of Green's function and its antiderivative. In contrast to the subsonic case, we note that even at $N = 3$ the Green function of the corresponding equation is a singular generalized function (a simple layer on a cone $z = m\|x\|$) (see Chap. 1). This is required for each N separate construction of the BIE, and rather subtle problems arise of determining convolutions of singular functions, such as layers on a cone and a cylinder.

References

1. Alexeyeva, L.A.: Boundary element method of boundary value problems of elastodynamics by stationary running loads. Int. J. Eng. Anal. Bound. Elem.T. (11), 37–44 (1998)
2. Alexeyeva, L.A.: Singular boundary integral equations of boundary value problems of elastodynamics in the case of subsonic running loads. Differ. Equ. **46**(4), 512–519 (2010)
3. Alexeyeva, L.A.: Singular boundary integral equations of boundary value problems in the theory of elasticity under supersonic transport loads. Differ. Equ. **53**(3), 327–342 (2017)
4. Vladimirov, V.S.: Equations of mathematical physics—Moskow "Nauka" (1981)

Chapter 6
Fundamental and Generalized Solutions of Maxwell Equations and Their Properties

In this chapter, using a unified methodology based on the generalized Fourier transform, solutions to Maxwell's equations are constructed in the space of generalized functions under the action of various currents and charges, nonstationary and stationary (harmonic in time). Periodic and non-periodic cases in the spatial variable that are important for applications are considered, for each of which the form of new Green's tensors for an infinite medium is presented in a form convenient for calculations.

6.1 Fundamental Solutions of Maxwell Equations

Let us consider the system of Maxwell equations for a homogeneous electromagnetic medium [1–3]:

$$\begin{cases} -\varepsilon \dfrac{\partial E}{\partial t} + rotH = j(x,t), \\ rotE + \mu \dfrac{\partial H}{\partial t} = 0, \end{cases} \tag{6.1}$$

$$\varepsilon \mathrm{div} E = \rho(x,t), \tag{6.2}$$

$$\mathrm{div} H = 0. \tag{6.3}$$

Here the electrical conductivity ε and magnetic permeability μ are constant, H, E are the magnetic and electric field intensities, $\rho(x,t)$ is electric charge density, and $j(x,t)$ is electric current density. It is required to determine H, E if the currents acting in the medium and the charge density are known.

L. Alexeyeva and A. Bayegizova, *Generalized Functions Method in Boundary Value Problems for Wave Equations*, Industrial and Applied Mathematics,
https://doi.org/10.1007/978-981-95-2802-8_6

Note that for given currents $j(x, t)$, , the system of Eqs. (6.1), taking into account the initial boundary conditions, is sufficient to determine H, E. Equation (6.3) is a consequence of the second equation of system (6.1) and is not independent. Equation (6.2) relates the acting charges to the electric field. From the first equation of system (6.1), taking into account (6.2), the well-known charge conservation equation follows:

$$\frac{\partial \rho}{\partial t} + \operatorname{div} j(x, t) = 0. \tag{6.4}$$

At given currents it is convenient to use it to determine the charge. Nonstationary system of Eqs. (6.1) of hyperbolic type, its characteristic equation is

$$n_t^2(n_t^2/c^2 - \|n\|^2)^2 = 0,$$

where speed of propagation of EM waves $c = 1/\sqrt{\mu\varepsilon}$, $\|n\| = \sqrt{n_1^2 + n_2^2 + n_3^2}$. That is, this is the cone of characteristic normals—classical light cone.

Next we will consider solutions of system (6.1) on the space of generalized vector functions of slow growth $S_6(R^4) = \{f(u) = (f_1(u), ..., f_6(u)), u \in R^4\}$.

As is known, solutions can suffer jumps on characteristic surfaces. Let us denote by F such a surface: $n = (n_1, n_2, n_2, n_t)$ is a unit normal to it in R^4. In coordinate space R^3, it corresponds to a wave front moving with speed c. Let us denote by $m = (m_1, m_2, m_3)$ the *wave vector*: $m_i = n_i/\|n\|$, $i = 1, 2, 3$;

then

$$c = -n_t/\|n\|. \tag{6.5}$$

If $\boldsymbol{U}$ is a classical solution (6.1), then, considered as a generalized solution on $S_6\prime(R^4)$, it will satisfy the following system of equations, written in matrix form:

$$Mu + G[u]\delta_F(x, t) = j(x, t), \tag{6.6}$$

$$u(x, t) = \begin{pmatrix} E(x, t) \\ H(x, t) \end{pmatrix}, \quad j(x, t) = \begin{pmatrix} j(x, t) \\ \mathbf{0} \end{pmatrix},$$

$$\mathrm{M} = \begin{pmatrix} -\varepsilon \mathrm{I}\partial_t & \Phi \\ \Phi & \mu \mathrm{I}\partial_t \end{pmatrix}, \quad \mathrm{G} = \begin{pmatrix} \varepsilon c \mathrm{I} & \Gamma \\ \Gamma & -\mu c \mathrm{I} \end{pmatrix},$$

$$\Phi = \begin{pmatrix} 0 & -\partial_z & \partial_y \\ \partial_z & 0 & -\partial_x \\ -\partial_y & \partial_x & 0 \end{pmatrix}, \quad \Gamma = \begin{pmatrix} 0 & -m_z & m_y \\ m_z & 0 & -m_x \\ -m_y & m_x & 0 \end{pmatrix}, \quad \mathbf{0} = \begin{pmatrix} 0 \\ 0 \\ 0 \end{pmatrix}.$$

Here SGF $\delta_F(x, t)$ is a simple layer on F, the density of which is determined through the jump

$$[u]_F = \lim_{\varepsilon \to +0} (u(x + \varepsilon m, t) - u(x - \varepsilon m, t)).$$

Therefore, in order for $U \in S_6'(R^4)$ to be a generalized solution (6.1), the following conditions must be met:

$$\mathrm{G}_{ij}\left[u_j\right]_F = 0. \tag{6.7}$$

On moving wave fronts they have the following form:

$$[E]_{F_t} = (\mu/\varepsilon)^{1/2}[H \times m]_{F_t}, \quad [m \times E]_{Ft} = (\mu/\varepsilon)^{1/2}[H]_{F_t}. \tag{6.7*}$$

If there are no field waves in front of the front, then at the front the three vectors E, H, m are mutually orthogonal, which indicates the transverse nature of the electromagnetic waves at the fronts.

Let us consider four classes of solutions to system (6.1), generated by currents of various types: nonstationary waves, nonstationary spatially periodic over z waves, monochromatic waves with frequency ω_0, and spatially periodic over z monochromatic waves. It is required to construct solutions of system (6.1) in the class of generalized functions in all these cases.

6.2 Nonstationary Solutions of Maxwell Equations: Green Tensor

Using the Fourier transform, we reduce system (6.1) to the form

$$\begin{pmatrix} -\varepsilon\omega \mathrm{I} & \Phi \\ \Phi & \mu\omega \mathrm{I} \end{pmatrix} \overline{U} = i \begin{pmatrix} \bar{j}(\xi, \omega) \\ \mathbf{0} \end{pmatrix}. \tag{6.8}$$

Here

$$\Phi = \begin{pmatrix} 0 & -\xi_3 & \xi_2 \\ \xi_3 & 0 & -\xi_1 \\ -\xi_2 & \xi_1 & 0 \end{pmatrix};$$

a bar means Fourier transform. Solving the system, we obtain the Fourier components of the fields

$$\begin{pmatrix} \overline{E}_1 \\ \overline{E}_2 \\ \overline{E}_3 \end{pmatrix} = \frac{-i}{\varepsilon\omega(k^2 - \|\xi\|^2)} \begin{pmatrix} k^2 - \xi_1^2 & -\xi_1\xi_2 & \xi_1\xi_3 \\ \xi_1\xi_2 & k^2 - \xi_2^2 & -\xi_2\xi_3 \\ -\xi_1\xi_3 & \xi_2\xi_3 & k^2 - \xi_3^2 \end{pmatrix} \cdot \begin{pmatrix} \bar{j}_1 \\ \bar{j}_2 \\ \bar{j}_3 \end{pmatrix},$$

$$\begin{pmatrix} \overline{H}_1 \\ \overline{H}_2 \\ \overline{H}_3 \end{pmatrix} = \frac{1}{(k^2 - \|\xi\|^2)} \begin{pmatrix} 0 & -\xi_3 & \xi_2 \\ \xi_3 & 0 & -\xi_1 \\ -\xi_2 & \xi_1 & 0 \end{pmatrix} \cdot \begin{pmatrix} \bar{j}_1 \\ \bar{j}_2 \\ \bar{j}_3 \end{pmatrix}, \tag{6.9}$$

where $k = \frac{\omega}{c}$, $\|\xi\| = \sqrt{\xi_1^2 + \xi_2^2 + \xi_3^2}$. The current density divergence, according to (6.4), is equal to

$$\sum_{k=1}^{3} \xi_k \bar{j}_k(\xi, \omega) = \omega \bar{\rho}(\xi, \omega). \tag{6.10}$$

We write formula (6.9) as follows:

$$\bar{u}_n = \overline{U}_{nk} \bar{j}^k, \quad n, k = \overline{1, 6}, \tag{6.11}$$

$$\{\overline{U}_{ik}\} = \begin{pmatrix} \frac{1}{\varepsilon}\overline{\mathrm{V}}^E & \overline{\mathrm{V}}^M \\ \overline{\mathrm{V}}^M & -\frac{1}{\mu}\overline{\mathrm{V}}^E \end{pmatrix}, \qquad \overline{\mathrm{V}}^M = \frac{1}{(k^2 - \|\xi\|^2)} \begin{pmatrix} 0 & -\xi_3 & \xi_2 \\ \xi_3 & 0 & -\xi_1 \\ -\xi_2 & \xi_1 & 0 \end{pmatrix},$$

$$\overline{\mathrm{V}}^E = \frac{1}{\omega\,(k^2 - \|\xi\|^2)} \begin{pmatrix} k^2 - \xi_1^2 & -\xi_1\xi_2 & \xi_1\xi_3 \\ \xi_1\xi_2 & k^2 - \xi_y^2 & -\xi_2\xi_3 \\ -\xi_1\xi_3 & \xi_2\xi_3 & k^2 - \xi_z^2 \end{pmatrix}$$

(summation from 1 to 6 is carried out everywhere over repeating indices). Fourier transform matrix of Green's tensor $\mathbf{U}(x, t)$, each column of which corresponds to fields generated by pulsed concentrated currents (including magnetic) of the form:

$$J_k(x, t) = \delta_k^{[m]} \delta(x) \delta(t), \quad m, k = \overline{1, 6}. \tag{6.12}$$

To construct a solution, we will use the inverse Fourier transform.

Theorem 6.1 *The nonstationary Green's tensor* $\mathbf{U}(x, t)$ *is a fundamental solution of equations* (6.1) *that satisfies the radiation conditions*:

$$\mathbf{U}(x, t) = 0 \quad \text{at}\, t < 0; \quad \mathbf{U}(x, t) = \mathbf{0} \text{ at } \|x\| > ct, \tag{6.13}$$

Has the Following Form:

$$\mathbf{U}(x, t) = \begin{pmatrix} \frac{1}{\varepsilon}\mathrm{V}^E(x, t) & \mathrm{V}^M(x, t) \\ \mathrm{V}^M(x, t) & -\frac{1}{\mu}\mathrm{V}^E(x, t) \end{pmatrix}, \tag{6.14}$$

$$V_{mn}^E = -\frac{\delta_{mn}}{4\pi c^2 r} \frac{\partial}{\partial t} \delta(t - r/c) + \frac{1}{4\pi} \frac{\partial^2}{\partial x_m \partial x_n} \left(\frac{H(t - r/c)}{r} \right),$$

$$V_{mn}^M = \frac{1}{4\pi} e_{mnl} \frac{\partial}{\partial x_l}\left(\frac{1}{r}\delta(t - r/c)\right).$$

Here $\|x\| = r$, $\delta(t - r/c)$ is a simple layer on light cone $K^+ = \{(x, t) : r = ct\}$, and e_{lmn} is the Levi–Civita pseudotensor.

Proof To construct $\mathbf{U}(x, t) = \{U_{nk}(x, t)\}$, we perform an inverse Fourier transform for $\overline{U}_{nk}(\xi, \omega)$. For this let's use the inverse Fourier transform of functions:

$$\mathrm{F}^{-1}\left[\frac{1}{(k + i0)^2 - \|\xi\|^2}\right] = -\frac{1}{4\pi r}\delta\left(t - \frac{r}{c}\right),$$

$$\mathrm{F}^{-1}\left[\frac{1}{(\omega + i0)}\frac{1}{\left((k + i0)^2 - \|\xi\|^2\right)}\right] = \frac{i}{4\pi R}\theta(t - \frac{r}{c}) \tag{6.15}$$

As a result, after calculation we obtain a provable statement. It is easy to see that the radiation conditions are met.

Note that Green's tensor includes terms of two types: a singular one with a carrier at the front of the electromagnetic wave—an expanding sphere of radius ct, and a regular one in a ball of the same radius (except for $\boldsymbol{x} = 0$). It describes, for a fixed m, a dipole field with a dipole moment in the direction of the axis X_m, which is established behind the front of the shock EM wave.

For given nonstationary currents, the E, M are represented in the form of a tensor-functional convolution of currents with Green's tensor:

$$u_n = U_{nk}(x, t) * j^k(x, t),$$

which is converted to the form

$$E_n(x, t) = -\frac{\delta_{nm}}{4\pi\varepsilon c^2 r}\delta(t - r/c) * j^m + \frac{1}{4\pi\varepsilon}\frac{\partial^2}{\partial x^n \partial x^m}\left(\frac{1}{r}H(t - r/c)\right) * j^m,$$

$$H_n(x, t) = \frac{1}{4\pi} e_{nlm}\frac{\partial}{x_l}\left(\frac{1}{r}\delta(t - r/c)\right) * j^m. \tag{6.16}$$

If we use formula (6.10) for the charge, then the equalities can be written more simply.

Theorem 6.2 *For any current $\boldsymbol{J}(x, t) \in S_6\prime(R_+^4)$, the components of the electromagnetic field are determined by the relations.*

$$E = -\mathrm{grad}\varphi - \frac{\partial A}{\partial t}, \quad H = \mu^{-1}\mathrm{rot}\, A,$$

where $\varphi = \frac{1}{4\pi\varepsilon}\left(\frac{1}{r}\delta\left(t - \frac{r}{c}\right) * \rho(x, t)\right)$, $A = \frac{\mu}{4\pi}\left(\frac{1}{R}\delta\left(t - \frac{r}{c}\right) * j(x, t)\right)$.

Proof From (6.9), by virtue (6.10), it follows that

$$\overline{E} = \frac{i\boldsymbol{\xi}}{\varepsilon(k^2 - \|\xi\|^2)}\overline{\rho} - \frac{ik}{c\varepsilon}\frac{1}{(k^2 - \|\xi\|^2)}\overline{j}, \tag{6.17}$$

$$\overline{H} = -i\boldsymbol{\xi} \times \frac{1}{k^2 - \|\xi\|^2}\overleftarrow{j}. \tag{6.18}$$

By using the property of Fourier transform of derivatives, we obtain E, H:

$$E(x,t) = -\frac{1}{4\pi\varepsilon}\operatorname{grad}\left(\frac{1}{r}\delta\left(t - \frac{r}{c}\right) * \rho\right) - \frac{\mu}{4\pi}\left(\frac{1}{r}\delta\left(t - \frac{r}{c}\right) * \frac{\partial j}{\partial t}\right)$$
$$H(x,t) = -\frac{1}{4\pi r}\delta\left(t - \frac{r}{c}\right) * \operatorname{rot}\boldsymbol{j}.$$

Taking into account the introduced scalar and vector potentials, we obtain the relations to be proved. Note that all convolutions exist due to the semi-boundedness of the supports in t and the boundedness of Green's tensor for a fixed t in x. For regular volumetric currents and charges, the formulas can be presented in integral form:

$$\varphi(x,t) = \frac{1}{4\pi\varepsilon}\int\limits_{r(x,y)<ct}\frac{1}{r(x,y)}\rho(r(x,y), t - \frac{r(x,y)}{c})dV(y),$$

$$A(x,t) = \frac{\mu}{4\pi}\int\limits_{r(x,y)<ct}\frac{1}{r(x,y)}\boldsymbol{j}(r(x,y), t - \frac{r(x,y)}{c})dV(y),$$

$$E = -\frac{1}{4\pi\varepsilon}\operatorname{grad}\int\limits_{r(x,y)<ct}\frac{x-y}{r(x,y)^3}dV(y)\int\limits_0^{t-r(x,y)/c} j(y,\tau)d\tau -$$
$$-\frac{\mu}{4\pi}\frac{\partial}{\partial t}\int\limits_{r(x,y)<ct}\frac{j(y, t - r(x,y)/c)}{r(x,y)}dV(y),$$

$$H = \frac{1}{4\pi}\int\limits_{r(x,y)<ct}\left[j(y, t - r(x,y)/c) \times \frac{x-y}{r(x,y)^3}\right]dV(y),$$

where $r(x,y) = \|x - y\|$, $dV(y) = dy_1dy_2dy_3$.

From them it follows that the electromagnetic field at point $\boldsymbol{x}$ at a fixed moment of time t forms currents in a ball of radius ct centered at this point.

For linear and surface currents, which can be represented as simple layers on the corresponding manifolds, to write these formulas in integral form, it is necessary to use the definition of convolution in the space of generalized functions.

6.3 Green Tensor of Stationary Maxwell Equations

Let us consider time-periodic solutions (6.1) for alternating currents of the form

$$j(x,t) = j(x)\exp(-i\omega_0 t), \tag{6.19}$$

where ω_0 is oscillation frequency, and $k = \frac{\omega_0}{c}$ is wave number. In this case, the solutions are also periodic with the same exponential factor. The corresponding system of equations for complex amplitudes has the form

$$\mathrm{M}_{mn}(\partial_1, \partial_2, \partial_3, -i\omega_0)u_n = j_m(x), \tag{6.20}$$

where $\mathrm{M}_{mn}(\partial_1, \partial_2, \partial_3, -i\omega_0)$ is differential operator M (6.6).

Theorem 6.3 Matrix W of fundamental solutions of Eq. (6.20), corresponding

$$j_n = \delta_n^{[m]}\delta(x),\, n, m = \overline{1,6}$$

and satisfying the Sommerfeld conditions at infinity has the following form:

$$\mathrm{W}(x) = \left\{W_n^m(x)\right\} = \begin{pmatrix} \frac{1}{\varepsilon}\mathrm{W}^E(x) & \mathrm{W}^M(x) \\ \mathrm{W}^M(x) & -\frac{1}{\mu}\mathrm{W}^E(x) \end{pmatrix},$$

where the elements of the matrices $\mathrm{W}^E(x)$ and $\mathrm{W}^M(x)$ are determined by the equalities:

$$W^E_{n[n]} = \frac{i\exp(ikr)}{4\pi cr}\left\{k + \frac{1}{kr}\left[(ik - 1/r)\left(1 - 3(x_n/r)^2\right) - (kx_n)^2/r\right]\right\},$$

$$W^E_{nm} = -\frac{ix_n x_m}{4\pi kcr^3}\exp(ikr)\left[\left(ik - \frac{2}{r}\right)\left(ik - \frac{1}{r}\right) + \frac{1}{r^2}\right], n \neq m,$$

$$W^M_{nm} = \frac{e_{lmn}x_l}{4\pi r^2}\exp(ikr)\left(ik - \frac{1}{r}\right), l, m, n = 1, 2, 3.$$

Proof Note that to determine it, in this case it is enough to invert the matrix $\overline{U}$ only with respect to $\boldsymbol{\xi}$, assuming $\omega = \omega_0$. Taking into account Sommerfeld radiation conditions,

$$\mathrm{F}_\xi^{-1}\left[\frac{1}{k^2 - \|\xi\|^2}\right] = -\frac{1}{4\pi}\frac{\exp(ikr)}{r}. \tag{6.21}$$

From (6.17), (6.18) we get the statement be proven.

For given stationary currents, complex amplitudes of EM field E, H are represented as a convolution over R only: $u_n = W_{nk} \underset{R}{*} j_k$. As follows from (6.4), in this case also $\rho(x, t) = \rho(x)\exp(-i\omega_0 t)$. This equality allows us to write the solution in a more convenient form.

Theorem 6.4 *For stationary currents of the form* (6.19) *satisfying the conditions.*

$$j(x) \in L_1(R^3), \tag{6.22}$$

the solution to system (6.20) *has the form*

$$\begin{aligned} E(x) &= -\frac{1}{4\pi\varepsilon}\operatorname{grad}\left(\rho(x) * \frac{\exp(ikr)}{r}\right) + \frac{ik}{4\pi\varepsilon c}\left(j(x) * \frac{\exp(ikr)}{r}\right), \\ H(x) &= \frac{1}{4\pi} rot\left(\frac{\exp(ikr)}{r} * j(x)\right). \end{aligned} \tag{6.23}$$

Proof follows from (6.17), (6.18). We invert them only by ξ and as a result we get what we need. Convolutions exist due to conditions (6.22) and the asymptotic properties of W for $r \to \infty$.

If currents are described by regular functions with support: supp$j(x) = D$,

$$\begin{aligned} E =& \frac{1}{4\pi\varepsilon}\int\limits_D \frac{(x-y)}{r(x,y)^2}\left(\frac{1}{r(x,y)} - ik\right)\rho(y)\exp\{-i\omega_0(t - r(x,y)/c)\}dV(y) + \\ &+ \frac{ik}{4\pi\varepsilon c}\int\limits_D \frac{\exp\{-i\omega_0(t - r(x,y)/c)\}}{r(x,y)} j(y)dV(y), \\ H =& \frac{1}{4\pi}\int\limits_D \left(\frac{1}{r(x,y)} - ik\right)\frac{[j(y) \times (x-y)]}{r(x,y)^2}\exp\{-i\omega_0(t - r(x,y)/c)\}dV(y) \end{aligned} \tag{6.24}$$

The first term here characterizes the so-called "near" field and is similar in structure to a static field. In the exponent, r/c indicates the retardation of the field at the point in question.

It is interesting that at charge density $\rho(x, t) = q\delta(x)\exp(-i\omega_0 t)$ passing to the limit $\omega_0 \to 0$ in (6.24) gives the Coulomb field of a static point charge q:

$$E(x) = \frac{q}{4\pi\varepsilon r^2}e_{\mathbf{x}}, \quad e_{\mathbf{x}} = \frac{x}{r}, \tag{6.25}$$

and the famous Biot–Savart–Laplace law for magnetostatics [2]:

$$H(x) = -\frac{1}{4\pi}\int_D \frac{j(y) \times (y - x)}{\|x - y\|^3} dV(y). \tag{6.26}$$

6.4 Stationary z-Periodic Solutions

Let us construct fields for spatially periodic alternating currents of the form: $\boldsymbol{j} = \boldsymbol{j}'(x, y)\exp\{-i(\omega_0 t - hz)\}$.

In particular, at $h = 0$ they correspond to currents and electromagnetic fields that are uniform along z (plane problem).

For determination of E and H in this case, their transformants should be reversed only in ξ_x, ξ_y, by $\xi_z = h$, $\omega = \omega_0$. Taking into account (p.204 [4])

$$\mathrm{F}_{\xi}^{-1}\left[(V^2 - \xi_x^2 - \xi_y^2)^{-1}\right] = -\frac{i}{4}H_0^{(1)}(Vr), \tag{6.27}$$

where $r = \sqrt{x^2 + y^2}$, $H_0^{(1)}$ is Hankel function of the first kind, and $V = \sqrt{k^2 - h^2}$, $\mathrm{Im}V \geq 0$. As a result, we obtain expressions for E, H:

$$\begin{pmatrix} E_x \\ E_y \\ E_z \end{pmatrix} = -\frac{1}{4\varepsilon\omega}\begin{pmatrix} k^2 + \frac{\partial^2}{\partial x^2} & \frac{\partial^2}{\partial x \partial y} & \frac{\partial^2}{\partial x \partial z} \\ \frac{\partial^2}{\partial y \partial x} & k^2 + \frac{\partial^2}{\partial y^2} & \frac{\partial^2}{\partial y \partial z} \\ \frac{\partial^2}{\partial x \partial z} & \frac{\partial^2}{\partial y \partial z} & k^2 + \frac{\partial^2}{\partial z^2} \end{pmatrix} H_0^{(1)}(Vr) \underset{x,y}{*} \begin{pmatrix} j_x \\ j_y \\ j_z \end{pmatrix},$$

$$\begin{pmatrix} H_x \\ H_y \\ H_z \end{pmatrix} = \frac{i}{4}\begin{pmatrix} 0 & -\frac{\partial}{\partial z} & \frac{\partial}{\partial y} \\ \frac{\partial}{\partial z} & 0 & -\frac{\partial}{\partial x} \\ -\frac{\partial}{\partial y} & \frac{\partial}{\partial x} & 0 \end{pmatrix} H_0^{(1)}(Vr) \underset{x,y}{*} \begin{pmatrix} j_x \\ j_y \\ j_z \end{pmatrix}.$$

Here the derivatives with respect to z correspond to the factor ih, and the designation $\underset{x,y}{*}$ is convolution with respect to x, y. In the case of regular volumetric currents, they can be written in integral form:

$$H = \frac{i\sqrt{k^2 - h^2}}{4}e^{i(hz - \omega_0 t)}\int_D \frac{[\boldsymbol{j}'(x', y') \times \boldsymbol{r}_0]}{\rho}H_1^{(1)}(\rho\sqrt{k^2 - h^2})dx'dy',$$

$$E = \frac{Vi}{4\varepsilon}e^{i(hz - \omega_0 t)}\int_D \left(H_1^{(1)}(Vr)\frac{\boldsymbol{r}_0}{\rho}\rho'(x', y') + \frac{ki}{c}H_0^{(1)}(V\rho)\boldsymbol{j}'(x', y')\right)dx'dy',$$

$$\rho = \sqrt{(x - x')^2 - (y - y')^2}, D = suppj'(x, y), \boldsymbol{r}_0 = \left(x - x', y - y', \frac{hr}{iV}\frac{H_0^{(1)}(Vr)}{H_1^{(1)}(Vr)}\right)$$

In particular, at $h = 0$ these formulas take the form:

$$H = \frac{ik}{4}e^{-i\omega_0 t}\int_D \frac{[\boldsymbol{j}'(x', y') \times \boldsymbol{r}]}{\rho}H_1^{(1)}(k\rho)dx'dy', \tag{6.28}$$

$$E = \frac{k}{4\varepsilon}e^{-i\omega_0 t}\int_D \left(i\rho'(x', y')H_1^{(1)}(k\rho)\frac{\boldsymbol{r}}{\rho} - \frac{1}{c}\boldsymbol{j}'(x', y')H_0^{(1)}(k\rho)\right)dx'dy', \tag{6.29}$$

$\boldsymbol{r} = \left(x - x', y - y'\right)$.

For linear and surface currents periodic in z, the formulas are similar; only the integrals are taken over the corresponding curves or surfaces. It is easy to see that the following theorem is true.

Theorem 6.5 *Matrix of fundamental solutions* $\left\{Q_{mn}\right\}$ *of system* (6.1), *corresponding to currents of the form:*

$$j_k = \delta_k^m \delta(x, y) \exp\{-i(\omega_0 t - hz)\}$$

and that satisfies the Sommerfeld radiation conditions has the form:

$$\begin{aligned} Q_{mn}^E &= -\frac{k}{4c} H_0^{(1)}(r\sqrt{k^2 - h^2})\delta_{mn} - \frac{1}{4\omega}\frac{\partial^2}{\partial x^m \partial x^n} H_0^{(1)}(r\sqrt{k^2 - h^2}), \\ Q_{nm}^M &= -\frac{i}{4} e_{mnl} \frac{\partial}{\partial x_l} H_0^{(1)}(r\sqrt{k^2 - h^2}), \quad m, n, l = 1, 2, 3. \end{aligned} \tag{6.30}$$

Here $(x_1, x_2, x_3) = (x, y, z)$, $r = \sqrt{x^2 + y^2}$, $\frac{\partial}{\partial x_3} \leftrightarrow ih$.

For given currents, complex amplitudes of EM field are represented as a convolution:

$u_n = Q_{nk} \underset{x,y}{*} j^k$.

Their integral recording is determined by the type of electric current and its carrier.

6.5 Nonstationary z-Periodic Solutions

Let us consider another type of field that is important for applications, generated by nonstationary currents periodic in z:

$$j = j(x, y, t)\theta(t)\exp(ihz). \tag{6.31}$$

Here $\theta(t)$ is Heaviside step function. At $h = 0, j_z = 0$ for example, they correspond to nonstationary currents that generate a nonstationary field uniform in z (plane problem). From (6.11), it follows that (we omit the exponential factor)

$$\begin{aligned} H_n &= e_{nkl} \frac{\partial}{\partial x_l} \left(F^{-1}\left[(k^2 - |\xi|^2)^{-1}\right]_{\xi_z = h}\right) \underset{x,y,t}{*} j^k, \quad E_n = A_{nk} \underset{x,y,t}{*} j^k, \\ A_{nk} &= \frac{\delta_{nk}}{\varepsilon c^2} \frac{\partial}{\partial t} F^{-1}\left[(k^2 - |\xi|^2)^{-1}\right]_{\xi_z = h} - \frac{i}{\varepsilon} \frac{\partial^2}{\partial x^n \partial x^k} F_{\xi,\omega}^{-1}\left[\omega^{-1}(k^2 - |\xi|^2)^{-1}\right]_{\xi_z = h} \end{aligned}$$

We use the formulas for incomplete inversion by ξ_x, ξ_y, ω (we omit their proof here):

$$F^{-1}_{\xi,\omega}\left[(k^2-|\xi|^2)^{-1}\right]_{\xi_z=h}=-\frac{c}{2\pi}\frac{\mathrm{Cosh}\sqrt{(ct)^2-r^2}}{\sqrt{(ct)^2-r^2}}H(ct-r),\; r=\sqrt{x^2+y^2},$$

$$F^{-1}_{\xi,\omega}\left[(\omega+i0)^{-1}(k^2-|\xi|^2)^{-1}\right]_{\xi_z=h}=\frac{i}{2\pi}\int\limits_0^{\sqrt{(ct)^2-r^2}}\frac{Coshs}{\sqrt{s^2+r^2}}ds.$$

Based on them it is easy to prove the theorem.

Theorem 6.6 *Matrix of complex amplitude* P(*x*,*y*,*t*) *of the fundamental solution of system* (6.1), *corresponding*

$$j^m=\delta^m_n\delta(x,y,t)\exp(ihz),\; n,m=\overline{1,6}$$

and satisfying the conditions, P(*x*,*y*,0) = *0 at r* ≠ *0,* P(*x*,*y*,*t*) = *0 at r* → ∞ *or t* → ∞,

$$P^E_{nk}=-\frac{\delta_{nk}}{2\pi c}\frac{\partial}{\partial t}\left(\frac{\mathrm{Cosh}\sqrt{(ct)^2-r^2}}{\sqrt{(ct)^2-r^2}}H(ct-r)\right)+\frac{1}{2\pi}\frac{\partial^2}{\partial x^n\partial x^k}\int\limits_0^{\sqrt{(ct)^2-\sigma^2}}\frac{\mathrm{Cosh}s}{\sqrt{s^2+r^2}}ds,$$

$$P^M_{kn}=\frac{c}{2\pi}e_{nkl}\frac{\partial}{\partial x_l}\left(\frac{\mathrm{Cosh}\sqrt{(ct)^2-r^2}}{\sqrt{(ct)^2-r^2}}H(ct-r)\right).$$

Here $\frac{\partial}{\partial z}\leftrightarrow ih(n,k,l=1,2,3)$*; the components of the complex amplitude matrix are defined similarly.*

At $h=0, j=j(x,t)H(t)$ from the above expressions, we get

$$P^E_{nk}=-\frac{\delta_{nk}}{2\pi c}\frac{\partial}{\partial t}\left(\frac{H(ct-r)}{\sqrt{(ct)^2-r^2}}\right)+\frac{1}{2\pi}\frac{\partial^2}{\partial x^n\partial x^k}\ln\left(\frac{ct}{r}+\sqrt{\left(\frac{ct}{r}\right)^2-1}\right),$$

$$P^M_{kn}=\frac{c}{2\pi}e_{nkl}\frac{\partial}{\partial x_l}\left(\frac{\theta(ct-r)}{\sqrt{(ct)^2-r^2}}\right).$$

For currents of the form (5.1), the field is represented as a convolution

$$U_n=P_{nk}(x,y,t)\underset{x,y,t}{*}j^k(x,y,t)\exp(ihz),$$

the integral representation of which is easy to write, similar to the above.

References

1. Tamm I.E.: Fundamentals of the theory of electricity. M.: Nauka, (1989)
2. Feynman R.P., Leighton R.B, Sands M.: The Feynman Lectures on Physics, v. 2: The Electromagnetic Field—Addison-Wesley, (1963)
3. Alekseeva L.A., Sautbekov S.S.: Fundamental solutions of Maxwell's equations, Differential equations. (1999). T. 35. No. 1.P. 125–127.
4. Vladimirov V.S.: Equations of mathematical physics—Moskow "Nauka", (1981)

Chapter 7
Generalized Solutions of Stationary Boundary Value Problems for Maxwell Equations

In this chapter, we construct the solutions of stationary boundary value problems for Maxwell equations which describe stationary oscillations of the EM field and EM waves with a constant oscillation frequency. This class of problems is very important for applications, as it makes it possible to study time-periodic processes in electromagnetic media.

Note that the apparatus of generalized functions allows one to naturally introduce representations for surface currents, charges, surface divergence, etc., the definition of which in classical electrodynamics does not satisfy the requirements of mathematical rigor.

7.1 Statement of Stationary Boundary Value Problems

We consider a monochromatic (*stationary*) electromagnetic field (E, H) of frequency ω in a limited area $S^- \in Œ$, generated by alternating electric and magnetic currents with a given density:

$$j^E = \{j_1, j_2, j_3\} \exp(-i\omega t), \; j^H = (j_4, j_5, j_6) \exp(-i\omega t).$$

It is assumed that the boundary surface S is closed, from the class of Lyapunov surfaces, or piecewise smooth, consisting of Lyapunov surfaces; $n(x)$ is the unit vector of the outer normal to S.

The amplitudes of the vectors E, H satisfy the system of stationary Maxwell equations:

$$i\varepsilon\omega E + rotH = j^E(x), \quad rotE - i\mu\omega H = j^H(x) \tag{7.1}$$

L. Alexeyeva and A. Bayegizova, *Generalized Functions Method in Boundary Value Problems for Wave Equations*, Industrial and Applied Mathematics,
https://doi.org/10.1007/978-981-95-2802-8_7

$j = \{j^E, j^H\} \in C(S+S^-) \cap C^1(S^-)$. The currents on the boundary surface are known, or

$$\text{BVP I} \qquad [n(x), E(x)] = j_S^H(x) \text{ at } x \in S, \tag{7.2}$$

or

$$\text{BVP II} \qquad [n(x), H(x)] = j_S^E(x) \text{ at } x \in S. \tag{7.3}$$

Here [a, b], (a, b) is a vector and scalar product. It is required to construct solutions to these boundary value problems (throughout we omit the time factor).

Comment. Often, instead of these boundary conditions, the tangent components of the field E or H on S are specified. It is easy to see that these conditions are equivalent (6.33, 6.34).

As is known (p. 44 [1]), a homogeneous boundary value problem ($\boldsymbol{j} = \boldsymbol{0}$) corresponding to the stated conditions has a unique zero solution, if $\text{Im}\varepsilon > 0$, $\text{Im}\mu > 0$. If they are real positive constants, then it can have non-zero solutions at some frequencies. These ω are called *resonant frequencies.*

For the densities of electric and magnetic charges ρ^E, ρ^H we have

$$i\omega\,\rho^E = \text{div}\, j^E, \quad i\omega\,\rho^H = \text{div}\, j^H. \tag{7.4}$$

This is the law of conservation of charge amplitudes for monochromatic fields, which we will use further for calculations.

The solution to the stated boundary value problems will be called classical. To determine them, we use BIEM. To construct the BIE, we will use GFM.

7.2 Statement of the Problem in the Space of Generalized Functions

Let us introduce the space of generalized vector functions D_6, defined on the space of infinitely differentiable finite functions $D_6(R^3) = \{(\phi_1(x), \ldots \phi_6(x)), \quad x \in R^3\}, \phi_k(x) \in D[2]$.

Let's denote $O_\varepsilon(x) = \{y : \|x - y\| < \varepsilon\}$, $\Gamma_\varepsilon(x) = \{y : \|x - y\| = \varepsilon\}$, according to the sign, $O_\varepsilon^\pm = O_\varepsilon(x) \cap S^\pm$, $\Gamma_\varepsilon^\pm = \Gamma_\varepsilon(x) \cap S^\pm$, $S^+ = R^3 - (S^- + S)$. Here we denote the characteristic function of the set S^- as

$$H_S^-(x) = \begin{cases} 1, & x \in S^-, \\ 0, & x \in S^+, \\ \alpha(x), & x \in S. \end{cases} \tag{7.5}$$

Here $\alpha(x) = \lim_{\varepsilon\to 0} V(O_\varepsilon^-(x))/V(O_\varepsilon(x))$, $V(...)$ is Euclidean volume of the specified set. It is easy to see that $4\pi\alpha(x) = \Omega(x)$ is the solid angle of the tangent to S the cone with its vertex at point x. If S is a smooth surface, then it coincides with the tangent plane and $\alpha(x) = 1/2$.

Using the rule of differentiation of generalized functions, we obtain a very useful equality for further calculations:

$$\partial_j(f(x)H_S^-(x)) = f_{,j}(x)\,H_S^-(x) - n_j f(x)\,\delta_S(x), \quad \forall f \in C^1(S^-) + C(S). \qquad (7.6)$$

Let us formulate the original boundary value problem in space $D_6'(R^3)$.

Let (E, H) be a classical solution to a boundary value problem. Let us introduce the functions $\hat{E} = H_S^-(x)E(x)$, $\hat{H} = H_S^-(x)H(x)$, similarly construct a continuation on R^3 for others, and we will consider them as generalized functions on $D_6'(R^3)$.

From (7.1) and (7.6) follow that $\{\hat{E}, \hat{H}\}$ satisfies in $D_6\prime(R^3)$ to the system

$$M(\partial_1, \partial_2, \partial_3)\begin{pmatrix}\hat{E}\\ \hat{H}\end{pmatrix} = \hat{j}(x) + \begin{pmatrix}[n,[H]]\\ [n,[E]]\end{pmatrix}\delta_S(x), \qquad (7.7)$$

$$\mathrm{M}(\partial_1, \partial_2, \partial_3) = \begin{pmatrix} i\varepsilon\omega I & G_0\\ G_0 & -i\mu\omega I\end{pmatrix}, \quad (G_0)_{nm} = -e_{nml}\partial_l \quad , \quad (n,\ m,\ l = 1,\ 2,\ 3).$$

Here in the vector product there are jumps on the surface:

$$[\hat{E}] = E^+ - E^- = -E|_S, \quad [\hat{H}] = H^+ - H^- = -H|_S.$$

It is a unit three-dimensional-matrix, and e_{nml} is antisymmetric unit Levi-Cevita pseudotensor: $[n, H]_l = e_{lmk}n_m H_k$. Everywhere over the same indices in the product, the summation is carried out from 1 to 3. Note that formula (6.38) is also true for a surface with an edge.

Equations for charges (7.4) in $D_6'(R^3)$ have the form:

$$\varepsilon div\hat{E} = \hat{\rho}^E + \varepsilon(n,[E])\delta_S(x), \quad -\mu div\hat{H} = \hat{\rho}^H - \mu(n,[H])\delta_S(x), \qquad (7.8)$$

where the first terms on the right correspond to volumetric charges, and the second ones describe surface electric and magnetic charges $\hat{\rho}_{S,}^E\ \hat{\rho}_{S.}^H$ Taking the divergence in (6.38), we obtain the charge conservation law in the space of generalized functions:

$$\begin{aligned} i\omega(\hat{\rho}^E + \varepsilon\,(n,[E])\delta_S(x)) &= h_S^-\,div\,j^E + (n,[j^E])\delta_S(x) + div([n,[H]]\,\delta_S(x),\\ i\omega(\hat{\rho}^H - \mu\,(n,[H])\delta_S(x)) &= h_S^-\,div\,j^H + (n,[j^H])\delta_S(x) + div([n,[E]]\,\delta_S(x). \end{aligned} \qquad (7.9)$$

Taking (7.8) into account, we obtain conditions for surface currents and charges:

$$div([n,[H]]\delta_S(x)) = (i\omega\varepsilon(n,[E]) - (n,[j^E]))\delta_S(x),$$

$$div([n,[E]]\delta_S(x)) = -(i\omega\mu(n,[H]) + (n,[j^H]))\delta_S(x). \tag{7.10}$$

Definition.
Let us call the surface rotor and the surface divergence of the vector E simple layers of the form.

$$rot_S E = [n,[E]]\delta_S(x), \quad div_S E = (n,[E])\delta_S(x).$$

Functions $rot_S H$, $rot_S E$ describe surface electric and magnetic currents on S, and functions $div_S E$, $div_S H$ describe surface electric and magnetic charges on S.
There are no magnetic charges or currents on the shielding surface

$$div_S H = 0, \quad rot_S E = 0, \tag{7.11}$$

or, in the absence of electrical charges and currents, there are

$$div_S E = 0, \quad rot_S H = 0. \tag{7.12}$$

From (7.10, 7.9) it follows that for this it is necessary that on S, respectively

$$div_S j^H = 0 \text{ or } div_S j^E = 0. \tag{7.13}$$

If we take into account the time multiplier, then the conservation law for surface charges can be written in a form similar to Maxwell relations:

$$\begin{aligned} &\frac{\partial \hat\rho_S^E}{\partial t} + div\,\hat j_S^E + div_S j^E = 0, \quad \hat\rho_S^E = \varepsilon\, div_S E, \quad \hat j_S^E = rot_S H \\ &\frac{\partial \hat\rho_S^H}{\partial t} + div\,\hat j_S^H + div_S j^H = 0, \quad \hat\rho_S^E = -\mu div_S H, \quad \hat j_S^H = rot_S E. \end{aligned} \tag{7.14}$$

Note that all the exact definitions of surface currents and charges introduced here through singular generalized functions correspond to their physical representations. In the physical literature (p. 180 [3]), "surface divergence" and "rotor" refer to the densities of the corresponding generalized functions introduced above, which denote $RotE = [n,[E]]$, $DivE = (n,[E])$.
Let us construct a solution (7.7) on $D_6\prime(R^3)$, which we will call *generalized*.

7.3 Generalized Solution of a Stationary Boundary Value Problem

The generalized solution (7.1) is constructed using Green's tensor $\hat U(x)$, which is a solution to Maxwell's equations:

$$\mathrm{M}(\partial_1, \partial_2, \partial_3)\hat{U}(x) = \mathrm{I}_6\, \delta(x). \tag{7.15}$$

It satisfies the Sommerfeld radiation conditions at infinity and has the form:

$$\hat{U}(x) = \{U_{ik}\} = \begin{pmatrix} \frac{1}{\varepsilon}V^E & V^M \\ V^M & -\frac{1}{\mu}V^E \end{pmatrix},$$

$$V^E = \frac{1}{\omega}\begin{pmatrix} k^2 + \partial_1^2\psi & \partial_1\partial_2\psi & -\partial_1\partial_3\psi \\ -\partial_1\partial_2\psi & k^2 + \partial_2^2\psi & \partial_2\partial_3\psi \\ \partial_1\partial_3\psi & -\partial_2\partial_3\psi & k^2 + \partial_3^2\psi \end{pmatrix}, V^M = -i\begin{pmatrix} 0 & \partial_3\psi & -\partial_2\psi \\ -\partial_3\psi & 0 & \partial_1\psi \\ \partial_2\psi & -\partial_1\psi & 0 \end{pmatrix},$$

where $\psi(x) = -(4\pi r)^{-1}\exp(ikr)$ is fundamental solution of the Helmholtz Eq. (1.28), $r = \|x\|$, $k = \omega/c$, $c = (\varepsilon\mu)^{-1/2}$.

Theorem 7.1. *If $j_S^E(x) \in L_1(S)$, $j_S^H(x) \in L_1(S)$ then the generalized solution of BVP can be represented in the form*

$$\hat{E} = E_0 + E^0 + \nabla\psi * [n, E]\delta_S + i\mu\omega\psi * [n, H]\,\delta_S - \nabla\psi * (n, E)\,\delta_S,$$

$$\hat{H} = H_0 + H^0 + \nabla\psi * [n, H]\delta_S - i\varepsilon\omega\psi * [n, E]\,\delta_S - \nabla\psi * (n, H)\,\delta_S;$$

$$E^0(x) = -i\omega\mu\,(\hat{j}^E * \psi) + \varepsilon^{-1}\nabla\psi * \hat{\rho}^E - [\nabla\psi * \hat{j}^H],$$
$$H^0(x) = i\omega\varepsilon\,(\hat{j}^H * \psi) - \mu^{-1}\nabla\psi * \hat{\rho}^H - [\nabla\psi * \hat{j}^E],$$

where (E_0, H_0) is a solution of homogeneous Eqs. (7.1).

Proof Using the property of Green's tensor, the solution of Eq. (7.7), taking into account (7.15), can be represented as next convolution

$$\begin{pmatrix} \hat{E} \\ \hat{H} \end{pmatrix} = \hat{U} * \hat{j}(x) - \hat{U} * \begin{pmatrix} [n, H] \\ [n, E] \end{pmatrix}\delta_S(x) + \begin{pmatrix} E_0 \\ H_0 \end{pmatrix}.$$

Using formulas for $\hat{U}$, we have

$$\begin{aligned}
&\hat{E} = E^0 - (i\varepsilon\omega)^{-1}\nabla\psi * (j^H, n)\delta_S - \\
&+ [\nabla\psi * [n, E]\delta_S] + i\mu\omega\psi * [n, H]\,\delta_S - (i\varepsilon\omega)^{-1}\Delta\psi * [n, H]\,\delta_S, \\
&\hat{H} = H^0 + (i\mu\omega)^{-1}\nabla\psi * (j^E, n)\delta_S + \\
&+ [\,\nabla\psi * [n, H]\delta_S] - i\varepsilon\omega\psi * [n, E]\,\delta_S + (i\mu\omega)^{-1}\Delta\psi * [n, E]\,\delta_S.
\end{aligned} \tag{7.16}$$

Here and below $[a * b]$, $(a * b)$ are the convolutions in the vector and scalar products of vectors $a\,(x)$ and $b(x)$:

$$[a * b]_l = e_{ljk} a_j(x) * b_k(x), \ \ (a * b) = a_j(x) * b_j(x).$$

The last terms in (7.16) can be transformed using (7.11, 7.12):

$$\Delta\psi * [n, H]\, \delta_S = \nabla\psi * div[n, H]\, \delta_S = -\nabla\psi * (i\omega\varepsilon(n, E) + (n, j^E))\delta_S,$$

$$\Delta\psi * [n, E]\, \delta_S = \nabla\psi * div[n, E]\, \delta_S = -\nabla\psi * (i\omega\mu(n, H) + (n, j^H))\delta_S.$$

Substituting into (7.16) and bringing similar terms, we obtain the statement being proved. It is easy to see that each convolution exists due to the boundedness of the support of one of its constituent functions and the conditions of the theorem.

For known $[n, E], \ [n, H]$ on S, the formulas of the theorem completely determine the solution of the problem in S^-.

Comment. Note that Theorem 7.1 is valid for the electromagnetic field in a limited region for any currents described by generalized vector functions, including singular ones.

7.4 Integral Representations of Stationary Solutions: Singular Boundary Integral Equations

To derive integral representations of the solution and the BIEqs, we use the integral representation of the characteristic function through the fundamental solution of the Helmholtz equation (see Lemma 1.5):

$$h_S^-(x) = k^2 \int_{S^-} \psi(r(x, y))\, dV(y) + \int_S \frac{\partial \psi(r(x, y))}{\partial n(y)}\, dS(y), \tag{7.17}$$

$r(x, y) = \|x - y\|$. The following lemma is true.

Lemma 7.1. *There is the equality:*
$\int_S [n(y), \nabla\psi(r(x, y))]\, dS(y) = 0$ *for* $\forall x$.
For $x \in S$ *singular integral, taken in the sense of the Principal Value.*

Proof Let $x \in S^-$. Then the integral in (7.17) is regular. Using the Ostrogradsky–Gauss formula, we obtain component-wise due to skew symmetry

$$e_{kjl} \int_S (n_l(y)\psi,_j (r(x, y))\; dS(y) = \; e_{kjl} \int_{S^-} \psi,_{jl} (r(x, y))\; dV(y) \; = 0.$$

Let $x \in S^+$ and $S^- \subset O_A(0)$. Then

$$e_{kjl}\int\limits_S n_l(y)\psi,_j(r(x,y))\,dS(y) =$$

$$= -e_{kjl}\int\limits_{\|y\|=A} n_l(y)\psi,_j(r(x,y))\,dS(y) + e_{kjl}\int\limits_{O_A(0)-S^-} \psi,_{jl}(r(x,y))dV(y) =$$

$$= -e_{kjl}\int\limits_{\|y\|<A} \psi,_{jl}(r(x,y))\,dS(y) = 0.$$

If $x \in S$ then $\psi'(r) \sim r^{-2}$ for $y \to x$. Therefore, the integral in the classical sense does not exist. Let's find its Principal Value.

Let $x \to x^* \in S,\;\; x \in S^-$, then

$$\lim_{x\to x^*} e_{kjl}\int\limits_{S_\varepsilon} n_l(y)\psi,_j(r)\,dS(y) + \lim_{\varepsilon\to 0}\psi'(\varepsilon)\int\limits_{\Gamma_\varepsilon^-} e_{kjl}e_l(x^*-y)e_j(x^*-y)\,dS(y) =$$

$$= \text{V.P.}\int\limits_{S_\varepsilon} e_{kjl}n_l(y)\psi,_j(r(x,y))\,dS(y) = 0,$$

$e(x-y)=(x-y)/r,\quad e_j = r,_j = (x_j-y_j)/r,\quad r = \|x-y\|$. The lemma is proven.

As follows from (7.16), to solve the boundary value problem it is necessary to determine (E, H) on the boundary surface S.

Theorem 7.2. *If the components of the vectors $j^E(x)$, $j^H(x)$ belong to the class $L_1(S^-+S)$ and S is Lyapunov's surface, then the solution of the stationary boundary value problem has an integral representation of the form:*

$$H_S^-(x)E(x) = E^0(x) + i\mu\omega\int\limits_S \psi(r(x,y))\left[n(y), H(y)\right]\,dS(y)+$$

$$+\,\text{V.P.}\int\limits_S \left\{(n(y),E(y))\nabla_y\psi(r) - \left[\nabla_y\psi(r),[n(y),E(y)]\right]\right\}dS(y),$$

$$H_S^-(x)H(x) = H^0(x) - i\varepsilon\omega\int\limits_S \psi(r)\left[n(y),E(y)\right]\,dS(y)+$$

$$+\,\text{V.P.}\int\limits_S \left\{\nabla_y\psi(r)\,(n(y),H(y)) - \left[\nabla_y\psi(r),[n(y),H(y)]\right]\right\}dS(y),$$

$$E^0(x) = -i\omega\mu\int\limits_{S^-}(j^E(y)\psi(r(x,y))dV(y)-$$

$$-\,(i\varepsilon\omega)^{-1}\nabla\int\limits_S \psi(r(x,y))(j^H(y),n(y))dS(y)+$$

$$+\varepsilon^{-1}\int\limits_{S^-}\nabla\psi(r(x,y))\rho^E(y)dV(y)-\mathbf{e}_l e_{lmn}\int\limits_{S^-}\partial_m\psi(r(x,y))j_n^H(y)dV(y),$$

$$H^0(x)=i\omega\varepsilon\int\limits_{S^-}(j^H(y)\psi(r(x,y))dV(y)+$$

$$+(i\mu\omega)^{-1}\nabla\int\limits_{S}\psi(r(x,y))(j^E(y),n(y))dS(y)-$$

$$-\mu^{-1}\int\limits_{S^-}\nabla\psi(r(x,y))\rho^H(y)dV(y)+\mathbf{e}_l e_{lmn}\int\limits_{S^-}\partial_m\psi(r(x,y))j_n^E(y)dV(y),$$

where $\nabla_y=(\partial_{y_1},\partial_{y_2},\partial_{y_3})$, $r=r(x,y)=\|x-y\|$.

Proof Let's consider Eq. (7.16). The integral notation of convolutions has the form specified in the theorem, where all volume and surface integrals exist for any $x\notin S$ due to the weak singularity of the integrands. For $x\in S^-$ this representation satisfies the stationary Maxwell Eqs. (7.1) in the classical sense by construction by virtue of (7.1) and the Du Bois-Reymond lemma (p. 95, [2]).

However, for $x\in S$ some of the integrals become divergent. Let us show that this representation is also preserved in this case if the divergent integrals are taken in the sense of Principal Value.

Describing the double vector product and rearranging the terms, for $x\in S^-$ we write the solution in the form

$$E(x)=E^0(x)+i\mu\omega\int\limits_{S}\psi(r)\left[n(y),H(y)\right]dS(y+$$

$$+\int\limits_{S}E(y)\frac{\partial\psi(r)}{\partial n(y)}dS(y)-\int\limits_{S}[E(y),[n(y),\nabla_y\psi(r)]]\,dS(y)),\tag{7.18}$$

$$H(x)=H^0(x)-i\varepsilon\omega\int\limits_{S}\psi(r)\left[n(y),E(y)\right]dS(y)+$$

$$+\int\limits_{S}H(y)\frac{\partial\psi}{\partial n(y)}dS(y)-\int\limits_{S}[H(y),[n(y),\nabla_y\psi(r)]]\,dS(y).\tag{7.19}$$

Let us consider the limit of each term in (7.18, 7.19) at $x\to x^*$, which we denote by E^k, H^k accordingly. It is easy to see that due to the weak singularity of the integrands

$$\lim_{x\to x^*}E^k(x)=E_l^k(x^*),\quad \lim_{x\to x^*}H^k(x)=H^k(x^*),\quad k=0,1.$$

As $\psi'(r)=1/(4\pi r^2)$, the integrals E^k, H^k $(k=2,3)$ diverge for $x\in S$. Let us use regularization (7.17) and Lemma 4.1. We calculate

$$H^2(x^*) = H(x^*) \lim_{x \to x^*} \int_S \frac{\partial \psi(r)}{\partial n(y)}\, dS(y)+$$

$$+ \lim_{x \to x^*} \int_S (H(y) - H(x^*)) \frac{\partial \psi(r)}{\partial n(y)} dS(y) = H(x^*)(1 - k^2 \int_{S^-} \psi(r) dV(y))+$$

$$+\text{V.P.} \int_S (H(y) - H(x^*)) \frac{\partial \psi(r)}{\partial n(y)} dS(y) = H(x^*)(1 - k^2 \int_{S^-} \psi(r) dV(y))+$$

$$+ \text{V.P.} \int_S (H(y) \frac{\partial \psi(r)}{\partial n(y)} dS(y) - H(x^*)\text{V.P.} \int_S \frac{\partial \psi(r)}{\partial n(y)} dS(y) =$$

$$= H(x^*)(1 - k^2 \int_{S^-} \psi(r) dV(y)) + \text{V.P.} \int_S (H(y) \frac{\partial \psi(r)}{\partial n(y)} dS(y)$$

$$- H(x^*)(0.5 - k^2 \int_{S^-} \psi(r) dV(y)) = 0,5H(x^*) + \text{V.P.} \int_S (H(y) \frac{\partial \psi(r)}{\partial n(y)} dS(y).$$

The proof is similar for $E^2(x^*)$. Let us consider

$$E^3 = \lim_{x \to x^*} \int_S [E(y),\, [n(y),\, \nabla \psi(r)]]\, dS(y) =$$

$$= \lim_{x \to x^*} \int_S [E(y) - E(x^*)\,,\, [n(y),\, \nabla \psi(r)]]\, dS(y) =$$

$$= \text{V.P.} \int_S [E(y),\, [n(y),\, \nabla \psi(r)]\,] dS(y) - [E(x^*),\, V.P. \int_S [n(y),\, \nabla \psi(r)]\,] dS(y) =$$

$$= \text{V.P.} \int_S [E(y),\, [n(y),\, \nabla \psi(r)]\,] dS(y).$$

By adding, rearranging, and bringing similar terms, we obtain the formulas of the theorem.

Remark 1. The proof of the theorems remains the same for external problems, only in this case it is sufficient to require that $j(x) = O(\|x\|^{-(1+\delta)})$, $\delta > 0$, for $\|x\| \to \infty$, the boundedness of the field on S, and the Sommerfeld radiation conditions at infinity:

$$(\partial E/\partial r - ikE) = O(1/r^2), \quad (\partial H/\partial r - ikH) = O(1/r^2) \text{ at } r \to \infty. \tag{7.20}$$

Remark 2. The theorems are also true for unbounded S. In this case, it is sufficient that $j(x) = O(r^{-(1+\delta)})$ for $\|x\| \to \infty$, the EM field is bounded on S and conditions (7.20).

To solve the posed boundary value problems, it is necessary to substitute known values of surface currents into the integral representation of solutions. In this case, it is easy to see that three BIEqs are separated from the system to determine the boundary values of H in the case of BVP I or E in the case of BVP II.

After solving the BIEqs using the formulas of Theorem 4.1, the solution is restored in the entire domain. Note that A. Poggio and E. Miller [4, 5] previously obtained these formulas in a different way.

Note also that the constructed BIEqs in the case of zero boundary conditions and volumetric currents give nonlinear relations for determining the resonant frequencies for internal problems. When constructing a discrete analogue of such a system with a numerical implementation, studying the behavior of the determinant of the corresponding system of linear algebraic equations depending on frequency allows them to be found.

References

1. Panich, O.I.: On the question of the solvability of external boundary value problems for the wave equation and Maxwell's equations. Adv. Math. Sci. **20**, 221–226 (1965)
2. Colton, D., Kress, R.: The impedance boundary value problem for the time harmonic Maxwell equations. Math. Meth. Appl. Sci. **3**, 475–487 (1981)
3. Tamm, I.E.: Fundamentals of the theory of electricity. M.: Nauka (1989)
4. Zakharov, E.V., Eremin, Yu.A.: On a method for solving axisymmetric problems of diffraction of electromagnetic waves on bodies of revolution. J. Comput. Math. Math. Phys. **19**(5), 1344–1348 (1979)
5. Alexeyeva, L.A.: Analogues of the Kirchhoff-Somigliana formulas in plane problems of elastodynamics. Appl. Math. Mech. **55**(2), 298–308 (1991)

Chapter 8
Generalized Solutions of Nonstationary Boundary Value Problems for Maxwell Equations

The study of the spread of nonstationary electromagnetic waves in limited areas leads to a number of difficulties associated with the occurrence of shock electromagnetic waves, which are characterized by a violation of the continuity of the field on the wave fronts. The latter prevents the use of the classic apparatus of mathematical physics or imposes increased requirements for the smoothness of electromagnetic fields, which significantly limits the class of solved tasks. The introduction of the concept of a generalized solution of Maxwell equations and the use of methods of generalized functions make it possible to cope with these difficulties. Moreover, the differentiation operation allows you to naturally derive the conditions on the wave fronts, which is a rather difficult task when using classical methods. When solving the BVPs by using GFM, it's possible to construct boundary integral equations and their solutions for shock EM waves.

8.1 Statement of Boundary Value Problems: Unity of Decisions

The nonstationary electromagnetic field (E, H) is considered in a closed limited area $S^- \subset R^3$, generated by variable electric and magnetics currents with given densities:

$$j^E(x,t) = \{j_1, j_2, j_3\},\ j^H(x,t) = (j_4, j_5, j_6),$$

$$j_k(x,t) \in L_1(D_+), \quad D_+ = \left\{(x,t),\ x \in (S^- + S),\ t \in [0,\infty)\right\}.$$

It is assumed that the boundary surface S is closed; from the class of Lyapunov's surfaces, $n(x)$ is a unit vector of the external normal to boundary S. Superficial currents are known at $x \in S$:

L. Alexeyeva and A. Bayegizova, *Generalized Functions Method in Boundary Value Problems for Wave Equations*, Industrial and Applied Mathematics,
https://doi.org/10.1007/978-981-95-2802-8_8

$$\text{BVP I } [n(x), E(x)] = j_S^H(x, t), \tag{8.1}$$

$$\text{BVP II } [n(x), H(x)] = j_S^E(x, t). \tag{8.2}$$

At the initial point in time are known

$$E(x, 0) = E_0(x), \quad H(x, 0) = H_0(x) \text{ at } x \in S^- + S \tag{8.3}$$

and the following approval conditions are fulfilled: at $x \in S$:

$$\text{BVP I } [n(x), E_0(x)] = j_S^H(x, 0) \tag{8.4}$$

$$\text{BVP II } [n(x), H_0(x)] = j_S^E(x, 0). \tag{8.5}$$

It is required to find the solutions to these BVPs.

Comment. Often, instead of these boundary conditions, the tangent components of the field $E(x, t)$ or $H(x, t)$ on S are known. It is easy to see that these conditions are equivalent.

It is required to determine the solutions to the Maxwell Eqs. (6.1), which we write here for components ($i, j, k = 1{,}2{,}3$):

$$\begin{aligned} &-\varepsilon \partial_t E_i(x, t) + e_{ijk} \partial_j H_k(x, t) = j_i^E(x, t), \\ &\mu \partial_t H_i(x, t) + e_{ijk} E_k(x, t) = j_i^H(x, t). \end{aligned} \tag{8.6}$$

It must satisfy the conditions for jumps on the fronts of shock electromagnetic waves:

$$\begin{aligned} &[E(x, t)]_{F_t} = (\mu/\varepsilon)^{1/2} [H(x, t) \times m(x, t)]_{F_t}, \\ &[m(x, t) \times E(x, t)]_{F_t} = (\mu/\varepsilon)^{1/2} [H(x, t)]_{F_t} \end{aligned} \tag{8.7}$$

and initial and boundary conditions.

Definition We call the solution to BVP *classical* if it is continuous and differentiated almost everywhere on D_+, perhaps, except the final number of wave fronts, on which it satisfies the conditions for jumps (8.7).

Here we introduce magnetic currents that are necessary to solve BVP. Note that the system of 6 Eqs. (8.6) is closed relative to 6 components of the vectors *E*, *H*. After their determination, the scalar equations of the Maxwell system (6.1) determine the charges.

8.2 Unity of Solutions of Nonstationary Boundary Value Problem

Let denote energy density $W(x,t) = 0,5(\varepsilon\|E\|^2 + \mu\|H\|^2)$, and Pointing vector $P(x,t) = [E, H]$.

Theorem 8.1 *If $u(x,t)$ is the classic solution to the BVP, then*

$$\int_{S^-} (W(x,t) - W(x,0))\, dV(x) =$$

$$= \int_0^t dt \int_S (n, [H, E])\, dV(x) + \int_0^t dt \int_{S^-} ((j^H, H) - (j^E, E))dV(x).$$

Proof Multiplying each Eq. (8.6) by E_i and H_i and summarizing over i, we get

$$\begin{aligned} &-0,5\varepsilon\, \partial_t\|E\|^2 + (e_{ijk}\partial_j H_k)E_i = (j^E, E),\\ &0,5\mu\, \partial_t\|H\|^2 + (e_{ijk}\partial_j E_k)H_i = (j^H, H). \end{aligned} \tag{8.8}$$

From these equalities, subtracting, we have

$$\partial_t W + e_{ijk}(H_i\partial_j E_k - E_i\partial_j H_k) = (j^H, H) - (j^E, E).$$

As $e_{ijk} = e_{kij} = -e_{kji}$, from equality it follows:

$$\partial_t W + \partial_j(e_{ijk}H_i E_k) = (j^H, H) - (j^E, E). \tag{8.9}$$

In the space of generalized functions, this equation has the form:

$$\begin{aligned} &\partial_t W + e_{ijk}\partial_j(H_i E_k) + \nu_t[W]|_F\delta_F(x,t) + e_{ijk}\nu_j[(H_i E_k)]|_F\delta_D(x,t) =\\ &= (j^H, H) - (j^E, E). \end{aligned} \tag{8.10}$$

Here $\nu = (\nu_1, \nu_2, \nu_3, \nu_t)$ is a single normal vector to characteristic surfaces. Integrating (8.10) in the region D_+, taking into account the jumps on the characteristic surfaces F (fronts F_t), and using the formula of the Ostrogradsky–Gauss in R^4, we receive *the law of EM field energy conservation*:

$$\begin{aligned} &\int_{S^-} (W(x,t) - W(x,0))\, dV(x) + e_{ijk}\int_0^t dt \int_S n_j H_i E_k\, dV(x) +\\ &\quad + \int_F (\nu_j e_{ijk}[H_i E_k]|_F + \nu_t[W(x,t)]|_F)dS(x,t)\\ &\quad = \int_0^t dt \int_{S^-} ((j^H, H) - (j^E, E))dV(x). \end{aligned} \tag{8.11}$$

Here $dS(x, t)$ is differential area of the characteristic surface F. In square brackets, there are the jumps of the corresponding expressions on F.

Let's consider a jump on the wave fronts in the third integral (8.11), which is denoted through g. Because

$[ab] = a^+b^+ - a^-b^- = a^+[b] - b^-[a]$,

taking into account the entered designations,

$$\begin{aligned} g(x,t) &= 0,5\,[\nu_t\mu\, H_iH_i + \nu_j e_{ijk} H_i E_k] + 0,5\,[\varepsilon\nu_t\, E_kE_k + \nu_j e_{ijk} H_i E_k] = \\ &= 0,5\,[H_i(\nu_t\mu\, H_i + \nu_j e_{ijk} E_k)] + 0,5\,[(\varepsilon\nu_t\, E_k + \nu_j e_{ijk} H_i)E_k] = \\ &= 0,5H_i^+\,[\nu_t\mu\, H_i + \nu_j e_{ijk} E_k] + 0,5\,[\varepsilon\nu_t\, E_k + \nu_j e_{ijk} H_i]E_k^+ + \\ &+ 0,5\,[H_i](\nu_t\mu\, H_i + \nu_j e_{ijk} E_k)^- + 0,5\,(\varepsilon\nu_t\, E_k + \nu_j e_{ijk} H_i)^-[E_k]. \end{aligned}$$

The first two terms, in virtue of (8.7), are zero. For the same reason, we get

$$\begin{aligned} g(x,t) &= 0,5\,[H_i](\nu_t\mu\, H_i + \nu_j e_{ijk} E_k)^- + 0,5\,(\varepsilon\nu_t\, E_k + \nu_j e_{ijk} H_i)^-[E_k] = \\ &= 0,5\,H_i^-\,(\nu_t\mu\,[H_i] + \nu_j e_{ijk}[E_k]) + 0,5\,(\varepsilon\nu_t\,[E_k] + \nu_j e_{ijk}[H_i])E_k^- = 0. \end{aligned} \tag{8.12}$$

From (8.11), taking into account the last equality, the approval of the theorem follows. The theorem is proven.

Corollary The rocking of energy on the fronts is described by the jump in the Pointing vector:

$$[W(x,t)]_{F_t} = c^{-1}(m(x,t), [P(x,t)]_{F_t}), \tag{8.13}$$

where $m(x, t)$ is wave vector.

The corollary of Theorem 8.1 is the unity of the BVP solution.

Theorem 8.2 *The classic solution to the first (second) boundary value problem of for t Maxwell Eqs.* (8.1) *is unity.*

Proof Suppose that there are two solutions, then their difference satisfies with zero initial data and regional conditions and $j^E = 0,\ j^H = 0$. Then

$$\begin{aligned} \int\limits_{S^-} W(x,t)\,dV(x) &= \int\limits_0^t dt \int\limits_S n_j e_{jkl} H_k E_l\,dV(x) = \\ &= \int\limits_0^t dt \int\limits_S ([n,H],E)\,dV(x) = -\int\limits_0^t dt \int\limits_S ([n,E],H)\,dV(x) = 0. \end{aligned}$$

Due to the positive $W(x,t) = 0,5(\varepsilon\|E\|^2 + \mu\|H\|^2) > 0$, we get $E \equiv 0,\ \ H \equiv 0$. The theorem is proven.

8.3 Statement of a Nonstationary BVP and Its Decision in Space $D_6'(R^4)$

Let's consider Maxwell equations in the space of generalized vector functions

$$D_6'(R^4) = \{\hat{f}(x,t) = (\hat{f}_1, ..., \hat{f}_6 \in D_6'(R^4)) : \hat{f}_j \in D'(R^4)\}.$$

Next, we use the designations:

$$D^- = S^- \times [0,\infty),\ D_+ = S \times [0,\infty),\quad H_D^-(x,t) = H_S^-(x)H(t).$$

$H_S^-(x)$ is characteristic function of S^-, and $\delta_D(x,t) = \delta_S(x)H(t)$ is the simple layer on D_+.

Let $(E(x,t), H(x,t))$ be a classical solution to a boundary value problem. Let us define it as a regular generalized vector function on the space of generalized vector functions $D_6'(R^4)$:

$$\hat{E}(x,t) = H_D^-(x,t)E(x,t),\quad \hat{H} = H_D^-(x,t)H(x,t),$$
$$\hat{j}(x,t) = H_D^-(x,t)j(x,t).$$

Using (8.1), we obtain

$$M(\partial_x, \partial_t)\begin{pmatrix}\hat{E}\\ \hat{H}\end{pmatrix} = \hat{j}(x,t) - \begin{pmatrix}[n,H]\\ [n,E]\end{pmatrix}\delta_D(x,t) + \begin{pmatrix}-\varepsilon E_0\\ \mu H_0\end{pmatrix} H_S^-(x)\delta(t). \quad (8.14)$$

Comparing the second terms on the right side with (8.1), we see that they describe surface currents, and the initial conditions work as volumetric pulse currents.

Using Green's tensor property, solution (8.14) can be represented as a convolution

$$\begin{pmatrix}\hat{E}(x,t)\\ \hat{H}(x,t)\end{pmatrix} = \mathbf{U}(x,t) * \hat{j}(x,t) - \mathbf{U} * \begin{pmatrix}[n(x), H(x,t)]\\ [n(x), E(x,t)]\end{pmatrix}$$
$$\delta_D(x,t) + \mathbf{U} \underset{x}{*} \begin{pmatrix}-\varepsilon E_0(x)\\ \mu H_0(x)\end{pmatrix} H_S^-(x). \quad (8.15)$$

For known densities of surface currents $[n,E]$, $[n,H]$ at S, these formulas completely determine the generalized solution of the problem.

To avoid cumbersome formulas in the integral representation of convolutions in (8.15), we next consider two boundary value problems in succession, which do not limit the generality due to the linearity of the original one.

8.4 Generalized Solution of Cauchy Problem

In this case, there is no boundary *S*; only the currents and initial conditions (8.3) are specified. Its generalized solution is given by the following theorem.

Theorem 8.3 *If* $(E_0(x), H_0(x)) \in D_6'(R^3)$, $(\hat{j}^E(x,t), \hat{j}^H(x,t)) \in D_6'(R^4)$ *then*

$$\hat{E}(x,t) = \mu\,\partial_t(\hat{j}^E * \psi) + \varepsilon^{-1}\nabla(\psi * \hat{\rho}^E) - [\nabla, \psi * \hat{j}^H] - c^{-2}(E_0 \underset{x}{*} \psi)_{,t} - \mu\Big[\nabla, [H_0 \underset{x}{*} \psi]\Big]6$$

$$\hat{H}(x,t) = -\varepsilon\partial_t(\hat{j}^H * \psi) - \mu^{-1}\nabla(\psi * \hat{\rho}^H) - [\nabla, \psi * \hat{j}^E] - c^{-2}(H_0 \underset{x}{*} \psi)_{,t} + \varepsilon\Big[\nabla, [E_0 \underset{x}{*} \psi]\Big]$$

$$\hat{\rho}^E(x,t) = \rho_0^E(x)H(t) - div(\hat{j}^E(x,t) \underset{t}{*} H(t)),$$

$$\hat{\rho}^H(x,t) = \rho_0^H H(t) - div(\hat{j}^H(x,t) \underset{t}{*} H(t)),$$

$$\rho_0^E(x) = \varepsilon div\, E_0(x), \quad \rho_0^H(x) = -\mu div\, H_0(x),$$

where $\psi(x,t) = -\frac{1}{4\pi r}\delta(t - t/c), \quad r = \|x\|.$

Proof In this case, the solution, as follows from (8.15), has the form:

$$\begin{pmatrix} \hat{E}(x,t \\ \hat{H}(x,t) \end{pmatrix} = \mathbf{U}(x,t) * \hat{j}(x,t) - \mathbf{U} * \begin{pmatrix} [n(x), H(x,t)] \\ [n(x), E(x,t)] \end{pmatrix}$$
$$\delta_D(x,t) + \mathbf{U} \underset{x}{*} \begin{pmatrix} -\varepsilon E_0 \\ \mu H_0 \end{pmatrix} H_S^-(x). \tag{8.16}$$

Using the properties of convolutions with the fundamental solution of the wave equation, we obtain the formula of the theorem if we take all convolutions componentwise and use the rules for its differentiation. All convolutions exist due to the semi-boundedness in *t* of the carriers of the functions included in them and the boundedness in *x* for a fixed *t* of the carrier of the wave function.

To write convolutions in integral form, it is important to know the smoothness properties of the initial field of currents. If the initial charge and currents are only continuous or piecewise continuous, and the conditions on the fronts (8.8) are satisfied, then the formulas of the theorem are equivalent to the following integral equalities:

$$
\begin{aligned}
4\pi E(x,t) = & -\mu\,\partial_t \int\limits_{r<ct} \frac{j^E(y,t-r/c)}{r(x,y)} dV(y) - \varepsilon^{-1}\nabla \int\limits_{r<ct} \frac{\rho^E(y,t-r/c)}{r(x,y)} dV(y) + \\
& + \left[\nabla, \int\limits_{r<ct} \frac{j^H(y,t-r/c)}{r(x,y)} dV(y)\right] + c^{-2}\partial_t \int\limits_{r=ct} \frac{E_0(y)}{r(x,y)} dS(y) \\
& + \mu\,[\nabla, \int\limits_{r=ct} \frac{H_0(y)}{r(x,y)} dS(y)] \\
4\pi H(x,t) = & \varepsilon\partial_t \int\limits_{r<ct} \frac{j^H(y,t-r/c)}{r(x,y)} dV(y) \\
& + \mu^{-1}\nabla \int\limits_{r<ct} \frac{\rho^H(y,t-r/c)}{r(x,y)} dV(y) + \\
& + \left[\nabla, \int\limits_{r<ct} \frac{j^E(y,t-r/c)}{r(x,y)} dV(y)\right] \\
& + c^{-2}\partial_t \int\limits_{r=ct} \frac{H_0(y)}{r(x,y)} dS(y) - \\
& - \varepsilon\left[\nabla, \int\limits_{r=ct} \frac{E_0(y)}{r(x,y)} dS(y)\right]
\end{aligned}
\tag{8.17}
$$

$$
\rho^E(x,t) = \rho_0^E(x)H(t) - div \int_0^t j^E(x,t-\tau)\ d\tau,
$$

$$
\rho^H(x,t) = \rho_0^H(x)H(t) - div \int_0^t j^H(x,t-\tau)\ d\tau.
$$

Here $r = \|x-y\| = r(x,y)$.

As an example, we consider the case of propagation of shock electromagnetic waves, which is useful for applications.

Let there be no currents ($j^E = 0,\ j^H = 0$). Let us assume that only in the region S^- the initial field is nonzero and $(E_0, H_0) \in C^2(S^- + S) \cup C^1(S^-)$. Let's find the field at $t > 0$. According to Theorem 8.3 in this case

$$
\begin{aligned}
\hat{\rho}_0^E(x,t) &= \varepsilon H(t)(div\ E_0(x) - (n(x), E_0)\delta_S), \\
\hat{\rho}_0^H &= -\mu H(t)(div\ H_0(x) - (n(x), H_0)\delta_S), \\
\hat{E}(x,t) = \varepsilon^{-1}\nabla(\psi * \hat{\rho}^E) &= \psi(x,t) * H(t)(\Delta E_0(x) - n(x)\text{div}E_0\delta_S - \varepsilon^{-1}\nabla((n, E_0)\delta_S)), \\
\hat{H}(x,t) = -\mu^{-1}\nabla(\psi * \hat{\rho}^H) &= \psi(x,t) * H(t)(\Delta H_0(x) - n(x)\text{div}H_0\delta_S + \mu^{-1}\nabla((n, H_0)\delta_S)).
\end{aligned}
$$

If the boundary S is a shock wave front at $t = 0$, then conditions (8.7) are satisfied on S, i.e.,

$$E_0(x) = (\mu/\varepsilon)^{1/2} H_0(x) \times n, \quad (\mu/\varepsilon)^{1/2} H_0(x) = n \times E_0(x), \quad x \in S. \tag{8.18}$$

It follows that on S there are $(E_0, n) = 0, \quad (H_0, n) = 0$.

That is, there are no surface charges. Moreover, according to the complete system of Maxwell equations, $\text{div} H_0 = 0$, because

$$\begin{aligned} \hat{E}(x,t) &= \psi(x,t) * H(t)(\Delta E_0(x) - n(x)\text{div}E_0(x)\delta_S(x)), \\ \hat{H}(x,t) &= \psi(x,t) * H(t)\Delta H_0(x), \\ \hat{\rho}_0^E(x,t) &= \varepsilon H(t) div\, E_0 = \rho_0^E H(t), \quad \hat{\rho}_0^H = 0. \end{aligned} \tag{8.19}$$

In the case of violation of the smoothness conditions of the initial conditions, it should be taken into account that when differentiating convolutions in the formulas of the theorem, simple layers will appear on the discontinuity surfaces of functions and their derivatives, where conditions (8.7) are not satisfied. They should be taken into account when taking the convolutions.

8.5 Integral Representation of the Solution to a Nonstationary BVP:Boundary Integral Equations

Let us consider the boundary value problem in a limited region with zero currents and initial conditions:

$$j^E = 0, \; j^H = 0, \quad E(x,0) = 0, \quad H(x,0) = 0, \quad x \in S^- + S. \tag{8.20}$$

In addition, the following conditions for matching the initial and boundary conditions are satisfied:

$$\text{BVP I} \quad j_S^H(x,0) = 0 \text{ at } x \in S, \tag{8.21}$$

$$\text{BVP II} \quad j_S^E(x,0) = 0 \text{ at } x \in S. \tag{8.22}$$

Its generalized solution is given by the following theorem.

Theorem 8.4 *Generalized solution of BVPs is representable in the form*

$$\hat{E}(x,t) = \left[\nabla\chi * \left[n, \dot{E}\right] \delta_D \right] - \mu\psi * [n, \dot{H}]\, \delta_D - \nabla\chi * (n, \dot{E})\, \delta_D$$

$$\hat{H}(x,t) = \left[\nabla\chi * \left[n, \dot{H}\right] \delta_D\right] + \varepsilon\psi * [n, \dot{E}] \delta_D - \nabla\chi * (n, \dot{H}) \delta_D$$

$$\hat{\rho}^E(x,t) = div([n, H] \delta_D) * H(t) = -\varepsilon (E, n) \delta_D$$

$$\hat{\rho}^H(x,t) = div([n, E] \delta_D) * H(t) = \mu(n, H) \delta_D$$

(everywhere a dot above a function means the derivative with respect to time t).

Proof From formulas (8.39) and (8.1) it follows that

$$\text{div}E = 0, \quad \text{div}H = 0. \tag{8.23}$$

We obtain a representation for charges from Theorem 8.1. By virtue of (8.39)–(8.41), (8.42), using the representation U, we have

$$\hat{E}(x,t) = [\nabla\psi * [n, E] \delta_D] - \mu\partial_t\psi * [n, H] \delta_D - \varepsilon^{-1}\Delta\chi * [n, H] \delta_D$$
$$\hat{H}(x,t) = [\nabla\psi * [n, H] \delta_D] + \varepsilon\partial_t\psi * [n, E] \delta_D + \mu^{-1}\Delta\chi * [n, E] \delta_D, \tag{8.24}$$

where $\nabla = (\partial_1, \partial_2, \partial_3) = \text{grad}$. The last terms can be transformed:

$$\Delta\chi * [n, H] \delta_D = \nabla\chi * div([n, H] \delta_D) = -\nabla\chi * (\varepsilon\partial_t(n, E)\delta_D)$$

$$\Delta\chi * [n, E] \delta_D = \nabla\chi * div([n, E] \delta_D) = \nabla\chi * (\mu\partial_t(n, H)\delta_D).$$

Substituting into (8.43), differentiating with respect to time, and bringing similar terms, we obtain

$$\hat{E}(x,t) = [\nabla\psi * [n, E] \delta_D] - \mu\psi * [n, \dot{H}] \delta_D - \nabla\psi * (n, E) \delta_D$$

$$\hat{H}(x,t) = [\nabla\psi * [n, H] \delta_D] + \varepsilon\psi * [n, \dot{E}] \delta_D - \nabla\psi * (n, H) \delta_D$$

because

$$[\nabla\psi * [n, H] \delta_D] = [\nabla\dot{\chi} * [n, H]\,] \delta_D = \left[\nabla\chi * \left[n, \dot{H}\right]\,\right] \delta_D + [\nabla\chi * [n, H_0] \delta_D]$$

$$-\nabla\psi * (n, H)\delta_D = -\nabla\dot{\chi} * (n, H) \delta_D = -\nabla\chi * (n, \dot{H}) \delta_D - \nabla\chi * (n, H_0) \delta_D$$

(the last terms in these equalities are equal to zero due to (8.39)). Hence the formulas of the theorem follow.

With known surface currents, Theorem 8.1 allows to determine the field inside the region. Integral representations of generalized solutions and boundary integral equations for determining unknown surface currents are given by the following theorem.

Theorem 8.5 *Classic solution by BVP is representable in the form*

$$4\pi H_D^- E(x,t) = \mu H(t) \int\limits_{S_t(x)} \frac{[n(y), \dot{H}(y, t-r/c)]}{r(x,y)} dS(y) -$$

$$-c^{-1} H(t) \int\limits_{S_t(x)} r^{-2}\left(\left[(x-y), [n(y), \dot{E}(y, t-r/c)]\right] - (x-y)\big(n(y), \dot{E}(y, t-\tau)\big)\right) dS(y) -$$

$$-H(t)\text{V.P.} \int\limits_{S_t(x)} r^{-3}\left(\left[(x-y), [n(y), E(y, t-r/c)]\right] - (x-y)\,(n(y), E(y, t-r/c))\right) dS(y)$$

$$4\pi H_D^- H(x,t) = -\varepsilon H(t) \int\limits_{S_t(x)} \frac{[n(y), \dot{E}(y, t-r/c)]}{r(x,y)} dS(y) -$$

$$-c^{-1} H(t) \int\limits_{S_t(x)} r^{-2}\left(\left[(x-y), [n(y), \dot{H}(y, t-r/c)]\right] - (x-y)\big(n(y), \dot{H}(y, t-\tau)\big)\right) dS(y) -$$

$$-H(t)V.P. \int\limits_{S_t(x)} r^{-3}\left(\left[(x-y), [n(y), H(y, t-r/c)]\right] - (x-y)(n(y), H(y, t-r/c))\right) dS(y).$$

Proof If to use formulas (8.29)–(8.35), we obtain the following integral representations of the convolutions in formulas of Theorem 8.1:

$$-\psi * [n, \dot{H}]\,\delta_D(x,t) = \frac{1}{4\pi} H(t) \int\limits_{S_t(x)} \frac{[n(y), \dot{H}(y, t-r/c)]}{r(x,y)} dS(y), \tag{8.25}$$

$$[\nabla\chi * [n, \dot{E}]\ \delta_D] = \left[\frac{x}{4\pi r^2}\left(\frac{H(t-r/c)}{r} + c^{-1}\delta(t-r/c)\right) * [n, \dot{E}]\ \delta_S(x) H(t)\right] =$$

$$= (4\pi)^{-1} H(t) \int\limits_0^t d\tau \int\limits_{S_\tau(x)} r^{-3}\left[(x-y), [n(y), \dot{E}(y, t-\tau)]\right] dS(y) -$$

$$-(4\pi c)^{-1} H(t) \int\limits_{S_t(x)} r^{-2}\left[(x-y), [n(y), \dot{E}(y, t-r/c)]\right] dS(y) -$$

$$-\nabla\chi * (n, \dot{E})\ \delta_D = (4\pi)^{-1} H(t) \int\limits_0^t d\tau \int\limits_{S_\tau(x)} (x-y)\, r^{-3}\big(n(y), \dot{E}(y, t-\tau)\big)\, dS(y) +$$

$$+(4\pi c)^{-1} H(t) \int\limits_{S_t(x)} r^{-2}(x-y)\big(n(y), \dot{E}(y, t-\tau)\big)\, dS(y). \tag{8.26}$$

In the double integrals of these convolutions, we change the order of integration and integrate over *t*. So, for example, the first integral (8.44), taking into account conditions (8.39), is transformed to the form

$$H(t) \int\limits_{S_t(x)} r^{-3} dS(y) \int\limits_{r/c}^{t} \left(\left[(x-y), [n(y), \dot{E}(y, t-\tau)]\right] - (x-y)\,\big(n(y), \dot{E}(y, t-\tau)\big)\right) d\tau =$$

$$= -H(t) \int\limits_{S_t(x)} r^{-3}\left([(x-y),[n(y),E(y,t-r/c)]] - (x-y)\,(n(y),E(y,t-r/c))\right) dS(y) =$$

$$= -H(t) \int\limits_{S_t(x)} r^{-3}\left([(x-y),[n(y),E(y,t-r/c)]] - (x-y)\,(n(y),E(y,t-r/c))\right) dS(y). \quad (8.27)$$

As a result, we obtain the indicated formulas.

Note that if $x \notin S$, then all integrands do not have singularities and the integrals exist. At $x \in S$, the first two integrals have a weak integrable singularity of order r^{-1}. The last integrals have strong non-integrable singularities at $y = x$ of order r^{-2}. Let us show that the formulas are also valid at the boundary if the integrals (8.47) are taken in the sense of the principal value and take into account the definition $h_S^-(x)$.

Let us introduce the notation $S^+ = R^3 - (S^- + S)$, $O_\varepsilon(x) = \{y : \ \|x-y\| < \varepsilon\}$, $\Gamma_\varepsilon(x) = \{y : \ \|x-y\| = \varepsilon\}$, $S_\varepsilon(x) = S - O_\varepsilon(x)$; these are entered according to the sign $S_\varepsilon^\pm(x) = S^\pm \cap O_\varepsilon^\pm(x)$, $O_\varepsilon^\pm = O_\varepsilon(x) \cap S^\pm$, $\Gamma_\varepsilon^\pm = \Gamma_\varepsilon(x) \cap S^\pm$. It is clear that the formulas of the theorem are true for interior points of the region bounded by any surface embedded in $S^- + S$. Let $x = x^* \in S$. We write these formulas for the region $S_\varepsilon^-(x^*)$. They have a similar form:

$$4\pi H^-_{S_\varepsilon + \Gamma_\varepsilon^-} E(x,t) = O(\varepsilon) + \mu \int\limits_{S_\varepsilon^-(x^*)} \frac{H(t-r/c)}{r(x,y)} [n(y), \dot{H}(y,t-r/c)]\, dS(y) -$$

$$-c^{-1} \int\limits_{S_\varepsilon^-(x^*)} \frac{H(t-r/c)}{r^2} \left([(x-y),[n(y),\dot{E}(y,t-r/c)]] - (x-y)(n(y),\dot{E}(y,t-\tau))\right) dS(y) -$$

$$- \int\limits_{S_\varepsilon^-(x^*)} \frac{H(t-r/c)}{r^3} \left([(x-y),[n(y),E(y,t-r/c)]] - (x-y)\,(n(y),E(y,t-r/c))\right) dS(y) +$$

$$+ \int\limits_{\Gamma_\varepsilon^-(x^*)} \frac{H(t-r/c)}{r^3} \left([(x-y),[n(y),E(x,t-r/c)]] - (x-y)\,(n(y),E(x,t-r/c))\right) dS(y).$$

Here we denote $O(\varepsilon)$ the first two integrals in formula of theorem 8.5 with respect to $\Gamma_\varepsilon^-(x^*)$.

Let $x = x*$, $t > 0$, then the left side of these formulas will vanish, since $x^* \notin S_\varepsilon^-(x^*)$. Now let's go to the limit at $\varepsilon \to 0$. The first two integrals tend to integrals over S due to the weak singularity of the integrand functions. The third integral tends to the integral in the sense of the principal value, which exists due to antisymmetry $(x * -y)/r$ and the order of singularity r^{-2}.

Let us consider in more detail the last integral, which we denote as I_4. Let's choose $\varepsilon < ct$. Then

$$I_4 = \frac{1}{\varepsilon^3} \int\limits_{\Gamma_\varepsilon^-(x^*)} \left(\left[(x^*-y),\left[n(y),E\left(x^*,t-\frac{\varepsilon}{c}\right)\right]\right] - (x^*-y)\left(n(y),E\left(x^*,t-\frac{\varepsilon}{c}\right)\right)\right) dS(y).$$

Let's move on to a spherical coordinate system with a center at point x^* and a vertical axis $n(x^*)$.

At $\Gamma_\varepsilon^-(x^*)$: $\|x^* - y\| = \varepsilon,\ n(y) = (x^* - y)/\varepsilon,\ dS(y) = \varepsilon^2 ds(e),\quad \|e = 1\|$, *ds(e)* is differential of the area of a unit sphere. Let's consider

$$\mathrm{I}_4 = \lim_{\varepsilon\to 0} \int\limits_{(\varepsilon e + x^*)\in\Gamma_\varepsilon^-(x^*)} \left(\left[e,\left[e, E(x^*, t - \varepsilon/c)\right]\right] - e\left(e, E(x^*, t - \varepsilon/c)\right)\right) ds(e) =$$

$$= \int\limits_{(e, n(x^*))<0} \left(\left[e,\left[e, E(x^*, t)\right]\right] - e\left(e, E(x^*, t)\right)\right) ds(e) = -2\pi E(x^*, t).$$

Transferring $2\pi E(x^*, t)$ to the left side, we obtain the assertion to be proved. The proof of the second formula of the theorem is similar. The theorem is proven.

Comment. It is easy to see that in the case of shock waves, the formulas, and their proofs are preserved for boundary points that do not fall on the wave front, because for such points, the smoothness conditions allow us to make the indicated passage to the limit. For points on the wave front, the strength values are not determined; we can only talk about the limiting values on different sides of the front.

Formulas of Theorem 8.6 on the boundary give boundary integral equations for solving the posed boundary value problems. To do this, one should substitute known values of surface currents (8.14) or (8.16) into the integral representation of solutions. In the case of the first boundary value problem, the BIEs for E are first solved. After determining E, the surface magnetic currents are determined, after which H can be found on the surface, and then the EM field in the region. For the second boundary value problem it is the other way around.

Note that the type of equation differs sharply from those for stationary BVPs. The obtained BIEqs in retarded potentials are of a non-classical type. They contain, in addition to the boundary values of intensities (surface currents), also their classical time derivatives. It is clear that for known currents, they are also known. However, the presence of such a derivative for unknown currents complicates the algorithm for the numerical solution of the BIEqs. It is required to introduce interpolation of time derivatives of the desired functions when solving the BIEqs step-by-step in time. This should be done using the values of the functions and their derivatives at previous time steps, starting from time zero, where the speed of the initial functions is given by the initial conditions.

In addition, the integration region depends on time and only at large times ($t > L/c$, L is diameter of the region S^-) coincides with S. The time variability of the integration region makes it possible to construct well-conditioned near-diagonal matrices of a resolving system of linear algebraic equations—a discrete analogue of BIEqs at numerical problem solving. To do this, you can move the initial moment of time, which allows to enter a limited number of contour points into the integration zone for any fixed boundary point. This simplifies the appearance of the system and increases the stability of numerical procedures.

Conclusion

As shown in the monograph, the method of generalized functions allows one to construct solutions to boundary value problems for linear equations and systems of partial differential equations of various types. It is especially effective for constructing solutions to equations and systems of hyperbolic and mixed type, where classical solution methods that require a certain smoothness of functions and their derivatives encounter great difficulties and limit the class of problems to be solved. The starting point in this method is the construction of Green's function for one equation or Green's tensor for tensor equations of mathematical physics and mechanics. Here we used the known fundamental solutions of the considered wave equations and constructed on their basis fundamental solutions of the system of Maxwell's equations. For arbitrary differential operators with constant coefficients, the question of constructing fundamental solutions was studied in detail by L. Hermander [51, 52]. For hyperbolic systems of equations of second order in derivatives, characteristic of problems in continuum mechanics and field theory, matrices (tensors) of fundamental solutions were constructed in [46]. To date, such fundamental solutions have been constructed for most equations and systems of mathematical physics and mechanics that describe the statics and dynamics of various media. The reader can find some of them in the bibliography for this monograph.

However, with the construction of solutions to boundary value problems for such systems of equations, especially systems of equations of hyperbolic and mixed type, the question is different. We talked a little about achievements in this area for solving boundary value problems of the theory of elasticity, thermoelasticity, and electrodynamics in the introduction. There is still a lot to be done here, for example, to construct Green's functions and tensors of nonstationary boundary value problems for areas with different boundary geometries, relevant for solving practically important problems in mechanics, physics, geophysics, seismology, earthquake-resistant construction, materials science and mechanical engineering, electrodynamics, and many others.

L. Alexeyeva and A. Bayegizova, *Generalized Functions Method in Boundary Value Problems for Wave Equations*, Industrial and Applied Mathematics,
https://doi.org/10.1007/978-981-95-2802-8

As shown in the monograph, the solution of such problems leads to a class of new singular boundary integral equations, not yet studied, for which it is necessary to develop a new theory, similar to that created by the works of many outstanding scientists for solving elliptic and parabolic boundary value problems.

This monograph was written based on the results of research carried out at the Institute of Mathematics and Mathematical Modeling of the Republic of Kazakhstan over the past 25 years with the financial support of the Science Committee of the Ministry of Science and Higher Education for republican programs of fundamental research in the field of physical and mathematical sciences.

This monograph was prepared and submitted for publication with the financial support of the Science Committee of the Republic of Kazakhstan under the republican program BR20281002 "Fundamental Research in Mathematics and Mathematical Modeling."

Bibliography

1. Gunther, N.M.: Potential theory and its application to the main problems of mathematical physics—Moskow "Gostehizdat", p. 283 (1953)
2. Bitsadze, A.V.: Boundary value problems for second order elliptic equations—Moskow "Nauka", p. 203 (1966)
3. Vekua, N.P.: Systems of singular integral equations—Moskow "Nauka", p. 379 (1970)
4. Muskhelishvili, N.I.: Singular integral equations—Moskow "Nauka", p. 512 (1968)
5. Kupradze, V.D.: Potential methods in the theory of elasticity—Moskow "Nauka", p. 472 (1963)
6. Mikhlin, S.G.: Course of mathematical physics—Moskow "Nauka", p. 684 (1968)
7. Perlin, P.I.: Boundary integral equations in the theory of elasticity—Moskow "Nauka", p. 312 (1977)
8. Jones, D.S.: Integral equations for the exterior acoustic problem. Q.J. Mech. Appl. Math. **27**, 129–142 (1974)
9. Colton, D.: The inverse scattering problem for acoustic waves. In: Dundee Conference on Ordinary and Partial Differential Equations. Springer-Verlag Lecture Notes in Mathematics, vol. 964, pp. 143–161 (1982)
10. Colton, D., Kress, R.: Iterative methods for solving the exterior Dirichlet problem for the Helmholtz equation with applications to the inverse scattering problem for low frequency acoustic waves. J. Math. Anal. Appl. **77**, 60–72 (1980)
11. Werner, P.: On the exterior boundary value problem of perfect reflection for stationary electromagnetic waves. J. Math. Anal. Appl. **7**, 348–396 (1963)
12. Werner, P.: On an integral equation in electromagnetic diffraction theory. J. Math. Anal. Appl. **14**, 445–462 (1966)
13. Panich, O.I.: On the question of the solvability of external boundary value problems for the wave equation and Maxwell's equations. Adv. Math. Sci. T. **20**, 221–226 (1965)
14. Zakharov, E.V., Eremin, Yu.A. On a method for solving axisymmetric problems of diffraction of electromagnetic waves on bodies of revolution. J. Comput. Math. Math. Phys. T. **19**(5), 1344–1348 (1979)
15. Colton, D., Kress, R.: The impedance boundary value problem for the time harmonic Maxwell equations. Math. Meth. Appl. Sci. **3**, 475–487 (1981)
16. Colton, D., Kress, R.: Methods of integral equations in scattering theory—Moskow "Nauka", p. 311 (1987)
17. Brebbia, K., Walker, S.: Application of boundary elements in technology—Moskow "Mir", p. 248 (1982)

L. Alexeyeva and A. Bayegizova, *Generalized Functions Method in Boundary Value Problems for Wave Equations*, Industrial and Applied Mathematics,
https://doi.org/10.1007/978-981-95-2802-8

18. Brebbia, C., Telles, J., Braubel, L.: Boundary element methods—Moskow "Mir", p. 528 (1987)
19. Beskos, D.E.: Boundary element method in dynamic analyses. Appl. mech. rev. **40**(1), 1–23 (1987)
20. Courant, R., Hilbert, D.: Methods of mathematical physics—Moskow-Leningrad "Gostekhiz-dat" vol. 1, 184 (1951)
21. Vladimirov, V.S.: Equations of mathematical physics—Moskow "Nauka" 512 (1981)
22. Kalmenov, T.Sh.: Boundary value problems for linear partial differential equations of hyperbolic type-Shymkent "Gylym" 328 (1993)
23. Aldashev, S.A.: Boundary value problems for multidimensional hyperbolic and mixed equations—Almaty "Gylym" 170 (1994)
24. Dzhumabaev, D.S., Asanova, A.T.: Unique solvability of boundary value problems with data on characteristics for systems of hyperbolic equations. Comput. Math. Math. Phys. **42**(11), 1673–1685 (2002)
25. Dzhumabaev, D.S., Asanova, A.T.: On the unique solvability of a nonlocal boundary value problem with data on intersecting lines for systems of hyperbolic equations//Differerential Equ.S **45**(3), 373–381 (2009)
26. Dzhumabaev, D.S., Asanova, A.T.: Correct solvability of nonlocal boundary value problems for systems of hyperbolic equations. Differerential Equ.S **41**(3), 337–346 (2005)
27. Asanova, A.T.: On a nonlocal boundary value problem for systems of quasilinear hyperbolic equations. Rep. Acad. Sci. **411**(1), 7–11 (2006)
28. Clifton, R.J.: Difference method in plane problems of dynamic elasticity. Mechanics (collection of translations) (1), 103–122 (1968)
29. Reker, V.V.: Numerical solutions of three-dimensional problems of the dynamic theory of elasticity. Appl. Mech. Ser. E. (1), 121–129 (1970)
30. Godunov, S.K., et al.: Numerical solution of multidimensional problems of gas dynamics—Moskow "Nauka" 400 (1976)
31. Yanenko, N.N.: Methods of fractional steps for solving multidimensional problems of mathematical physics. Novosibirsk: Publishing House. NSU, 195 (1967)
32. Tarabrin, G.T.: Application of the bicharacteristics method for solving nonstationary problems in the dynamics of anisotropic massifs. Struct. Mech. Des. Struct. (4), 38–43 (1981)
33. Baiteliev, T.B., Mardonov, B.: Solution of a plane dynamic problem of the theory of elasticity for a strip by the method of spatial characteristics. In the book: Problematic issues of rock mechanics.—Alma-Ata 194–201 (1972)
34. Alexeyeva, L.A.: Analogues of the Kirchhoff-Somigliana formulas in plane problems of elastodynamics. Appl. Math. Mech. **55**(2), 298–308 (1991)
35. Alexeyeva, L.A.: Boundary element method of boundary value problems of Elastodynamics by stationary running loads. Int. J. Eng. Anal. Bound. Elem. T (11), 37–44 (1998)
36. Alexeyeva, L.A., Sautbekov, S.S.: Method of generalized functions in stationary boundary value problems for Maxwell's equations. J. Comput. Math. Math. Phys. **40**(4), 611–622 (2000)
37. Alexeyeva, L.A.: Dynamic analogues of the Green and Gauss formulas for solutions of the wave equation in. Differ. Equ.S **31**(11), 1951–1953 (1995)
38. Alexeyeva, L.A.: Boundary integral equations of the initial-boundary value problem for the wave equation in. Differ. Equ.S 28(8), 1451–1453 (1992)
39. Alexeyeva, L.A.: Generalized solutions of nonstationary boundary value problems for Maxwell's equations. J. Comput. Math. Math. Phys. **42**(1), 76–88 (2002)
40. Alexeyeva, L.A.: Method of generalized functions in nonstationary boundary value problems for the wave equation. Math. J. **6**(1) (19), 16–32 (2006)
41. Alexeyeva, L.A.: Boundary integral equations of nonstationary BVP for wave equations. Int. Congress of Mathematicians. Abstracts. Madrid 436 (2006)
42. Dynamic analogues of the Somilyana formula for nonstationary dynamics of elastic media with an arbitrary degree of anisotropy. Appl. Math. Mech. **58**(2), 170–175 (1994)
43. Alexeyeva, L.A., Dildabayev, Sh.A., Zakiryanova, G.K., Zhanbyrbaev, A.B.: Boundary integral equations method in two-and three-dimensional problems of elastodynamics. Comput. Mech. **18**, 147–157 (1996)

44. Alexeyeva, L.A., Zakiryanova, G.K.: Generalized solutions of boundary value problems of dynamics of anisotropic elastic media. J. Mech. Behav. Mater. **16**(4–5), 259–267 (2005)
45. Alexeyeva, L.A., Kupesova, B.N.: Method of generalized functions in boundary value problems of coupled thermoelastodynamics. Appl. Math. Mech. **65**(2), 334–345 (2001)
46. Alexeyeva, L.A., Zakiryanova, G.K.: Green's matrix for strictly hyperbolic systems with second-order derivatives. Differ. Equ.S **37**(4), 488–494 (2001)
47. Alexeyeva, L.A., Zakiryanova, G.K.: Generalized solutions of initial-boundary value problems for second-order hyperbolic systems. J. Comput. Math. Math. Phys. **51**(7), 1280–1293 (2011)
48. Tamm, I.E.: Fundamentals of the theory of electricity. M.: Nauka (1989)
49. Feynman, R.P., Leighton, R.B, Sands, M.: The Feynman lectures on physics, vol. 2. The Electromagnetic Field—Addison-Wesley (1963)
50. Schwartz, L.: Mathematical methods for physical sciences—Moskow "Mir" (1965)
51. Hörmander, L.: Linear differential operators with partial derivatives—Moskow "Mir" (1965)
52. Hermander, L.: Analysis of linear partial differential operators. In: 4 volumes—Moskow "Mir" (1986)
53. Vladimirov, V.S.: Generalized functions in mathematical physics—Moskow "Nauka" (1979)
54. Vladimirov, V.S.: Equations of mathematical physics—Moskow "Nauka" (1981)
55. Vladimirov, V.S., Mikhailov, V.P., Vasharin, A.A., Karimova, Kh.H., Sidorov, Yu.V., Shabunin, M.I.: Collection of problems on the equations of mathematical physics—Moskow "Nauka" (1989)
56. Polyanin, A.D.: Handbook of linear equations of mathematical physics—Moskow "FIZ-MATLIT" (2001)
57. Special Functions Reference. Abramowitz, M., Stigan (eds.), I.—Moskow "Nauka" (1979)
58. Sommerfeld, A.: Die Greensche Funktion der Schwingungsglerchung. Jber. Deutsch Math. Verein, Bd 21, 309–353 (1912)
59. Atkinson, F.V.: On Sommerfeld's «Radiation condition». Philos Mag. **40**, 645–651 (1949)
60. Kleinman, R.E., Wendland, W.: On Neumann's method for the exterior Neumann problem for the Helmholtz equation. J Math. Anal. Appl. **57**, 170–202 (1977)
61. Colton, D., Kleinman, R.E.: The direct and inverse scattering problems for an arbitrary cylinder. Dirichlet boundary conditions. Proc. Royal Soc. Edinburgh, **86A**, 29–42 (1980)
62. Kyurkchan, A.G.: On one scattering problem for the Helmholtz equation. Reports of the USSR Acad. Sci. **275**(1), 48–51 (1984)
63. Bogolyubov, N.N., Logunov, A.A., Todorov, I.T.: Fundamentals of the axiomatic approach in quantum field theory—Moskow "Nauka" (1970)
64. Bogolyubov, N.N., Shirkov, D.V.: Introduction to the theory of quantized fields—Moskow "Nauka" (1973)
65. Landau, L.D., Lifshits, E.M.: Field theory—Moskow "Nauka" (1973)
66. Lax, P., Phillips, R.: Scattering theory—Moskow "Mir" (1971)
67. Petrovsky, I.G.: Partial differential equations—Moskow "FIZMATGIZ" (1972)
68. Hadamard, J.: Le probleme de Cauchy et les equations aux derivees partielles lineaires hyperboliques.—Paris (1932)
69. Alexeyeva, L.A.: Singular boundary integral equations of boundary value problems of elastodynamics in the case of subsonic running loads. Differ. Equ.S. **46**(4):512–519 (2010)
70. Alexeyeva, L.A.: Singular boundary integral equations of boundary value problems in the theory of elasticity under supersonic transport loads. Differ. Equ.S **53**(3), 327–342 (2017)
71. Bayegizova, A.S.: On one approach to solving the inverse problem for the wave equation. Vestnik KazNU 24–33 (2000)
72. Bayegizova, A.S.: On the functional dependence of amplitude with wave functions. Vestnik KazNU 31–42 (2000)
73. Bayegizova, A.S.: About one connection between amplitude and wave functions. Bulletin of KarSU 28–36 (2001)
74. Alexeyeva, L. A.: Generalized solutions of boundary value problems for one class of traveling solutions of the wave equation. Math. J. T. **8**(2) (30), 1–19 (2008)

75. Alexeyeva, L. A.: General functions method in transport boundary value problems of elasticity theory. IntechOpen. In: Book “Differential equations. Theory and current researches”. Chapter 8. pp. 129–161 (2018)
76. Alekseeva, L. A., Sautbekov, S. S.: Fundamental solutions of Maxwell’s equations. Differ. Equ.S. T. **35**(1), 125–127 (1999)
77. Alekseeva, L. A., Sautbekov, S. S.: Method of generalized functions in stationary boundary value problems for Maxwell’s equations. J. Comput. Math. Math. Physics. T. **40**(4), 611–622 (2000)
78. Alekseeva, L. A.: Generalized Solutions to time-dependent boundary value problems for Maxwell’s equations. Comput. Math. Math. Physic **42**(1), 76–88 (2002)

MIX
Papier aus verantwortungsvollen Quellen
Paper from responsible sources
FSC® C105338

If you have any concerns about our products,
you can contact us on
ProductSafety@springernature.com

In case Publisher is established outside the EU,
the EU authorized representative is:
Springer Nature Customer Service Center GmbH
Europaplatz 3, 69115 Heidelberg, Germany

Printed by Libri Plureos GmbH
in Hamburg, Germany